FLORA OF THE
BRITISH ISLES

ILLUSTRATIONS

A. R. CLAPHAM   T. G. TUTIN   E. F. WARBURG

FLORA OF THE BRITISH ISLES

ILLUSTRATIONS

PART II
ROSACEAE–POLEMONIACEAE

DRAWINGS BY
SYBIL J. ROLES

CAMBRIDGE
AT THE UNIVERSITY PRESS
1960

PUBLISHED BY

THE SYNDICS OF THE CAMBRIDGE UNIVERSITY PRESS

Bentley House, 200 Euston Road, London, N.W. 1
American Branch: 32 East 57th Street, New York 22, N.Y.

©

CAMBRIDGE UNIVERSITY PRESS

1960

Printed in Great Britain at the University Press, Cambridge
(Brooke Crutchley, University Printer)

PREFACE

The second part of these Illustrations of British plants follows the same general pattern as the first.

The intention once again is to provide a visual impression of the habit, and a selection of the chief features of the plants so as to assist with the appreciation of the technical descriptions given in *Flora of the British Isles* and the *Excursion Flora*.

The drawings have almost all been made from fresh specimens and are reproduced on a rather larger scale than that of the illustrations accompanying Bentham and Hooker's well-known *Handbook*. It is hoped that these features, combined with the grouping of illustrations to facilitate as far as possible easy comparison of related species will make them generally useful.

We should like to express once again our gratitude to Miss Roles for her patience and skill in the almost impossible task of providing small-scale drawings of 'typical' specimens.

We are greatly indebted to the following for supplying many of the plants illustrated in this Part: Miss J. Allison, Miss M. Atkins, P. W. Ball, P. R. Bell, M. Borrill, Miss A. P. Conolly, R. W. David, Miss E. W. Davies, Miss U. K. Duncan, J. R. S. Fincham, H. Gilbert-Carter, the late R. A. Graham, G. Halliday, M. K. Hanson, Miss J. M. Hartshorn, Miss J. E. Hibberd, E. K. Horwood, H. M. Hurst, A. C. Jermy, J. E. Lousley, D. McClintock, Miss P. A. Padmore, C. D. Piggot, M. E. D. Poore, T. E. D. Poore, C. E. Raven, J. E. Raven, N. W. Simmons, F. A. Sowter, Mrs F. le Sueur, F. J. Taylor, Mrs F. J. Taylor, Miss E. M. Thomas, D. H. Valentine, S. M. Walters, Miss M. McCullum Webster, W. T. Williams, and P. F. Yeo.

A.R.C.
T.G.T.
E.F.W.

PART II

ROSACEAE—POLEMONIACEAE

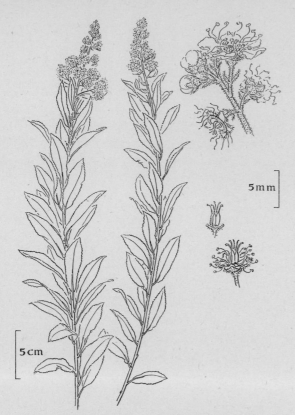

553. *Spiraea salicifolia* L. Willow Spiraea Pink

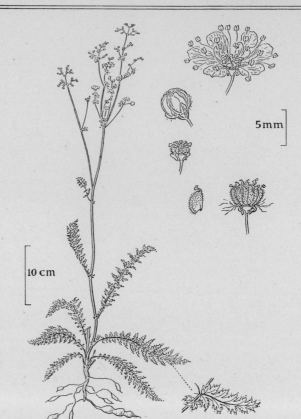

554. *Filipendula vulgaris* Moench Dropwort White

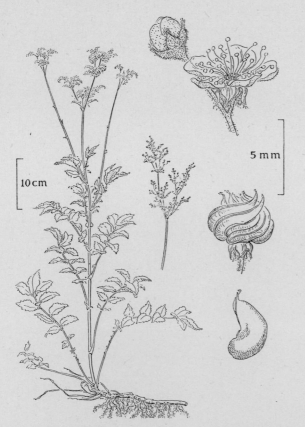

555. *Filipendula ulmaria* (L.) Maxim. Meadow-sweet White

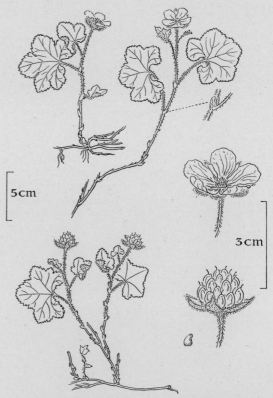

556. *Rubus chamaemorus* L. Cloudberry White

557. *Rubus saxatilis* L. Stone Bramble White

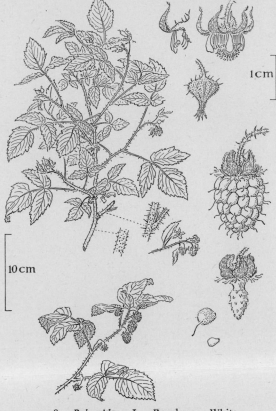

558. *Rubus Idaeus* L. Raspberry White

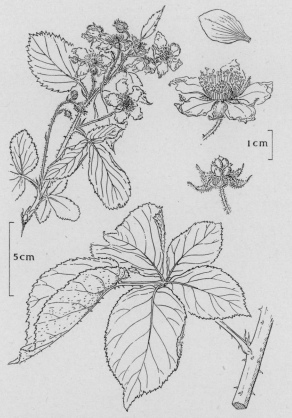

559. *Rubus nessensis* W. Hall Blackberry White

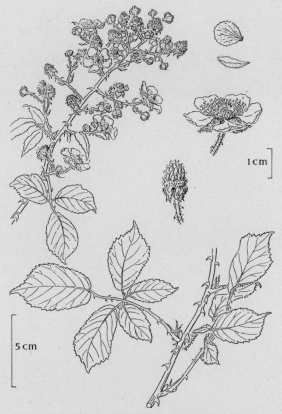

560. *Rubus ulmifolius* Schott Blackberry Pink

I-2

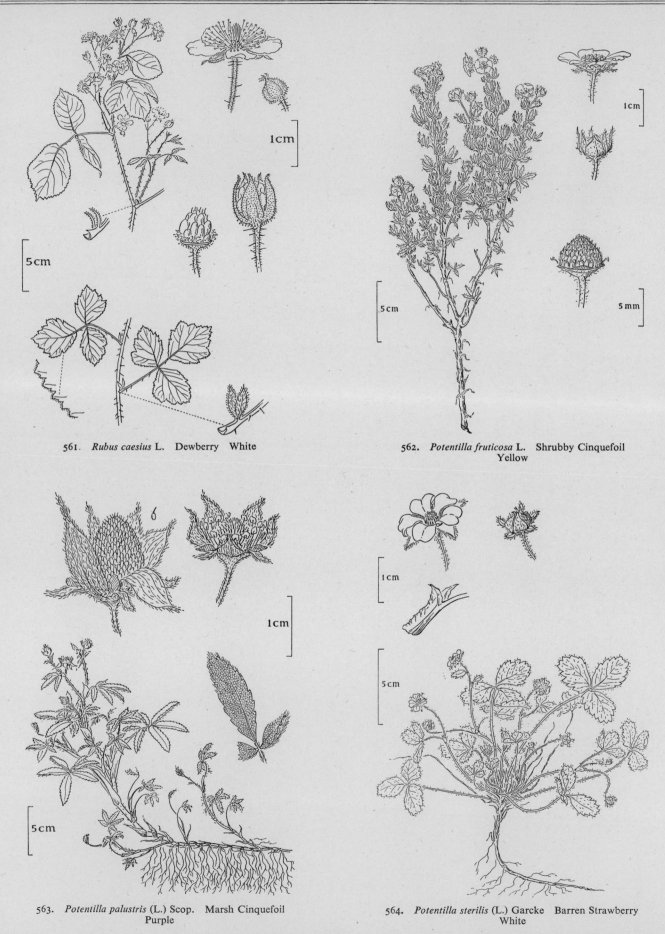

561. *Rubus caesius* L.   Dewberry   White

562. *Potentilla fruticosa* L.   Shrubby Cinquefoil
Yellow

563. *Potentilla palustris* (L.) Scop.   Marsh Cinquefoil
Purple

564. *Potentilla sterilis* (L.) Garcke   Barren Strawberry
White

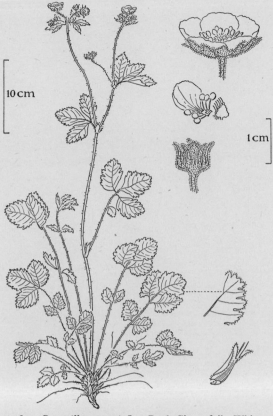

565. *Potentilla rupestris* L. Rock Cinquefoil White

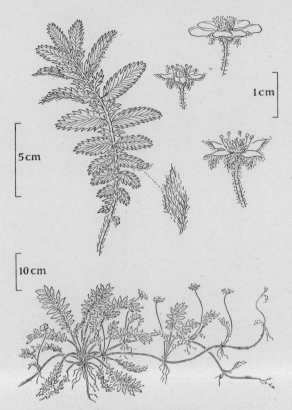

566. *Potentilla anserina* L. Silverweed Yellow

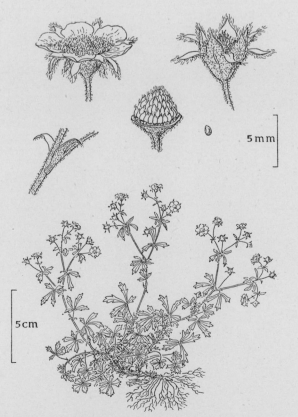

567. *Potentilla argentea* L. Hoary Cinquefoil Yellow

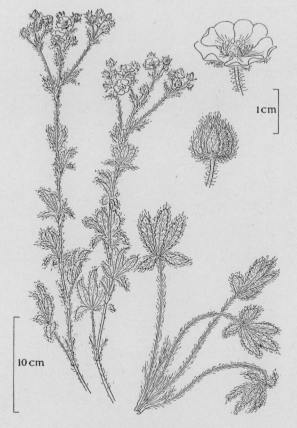

568. *Potentilla recta* L. Yellow

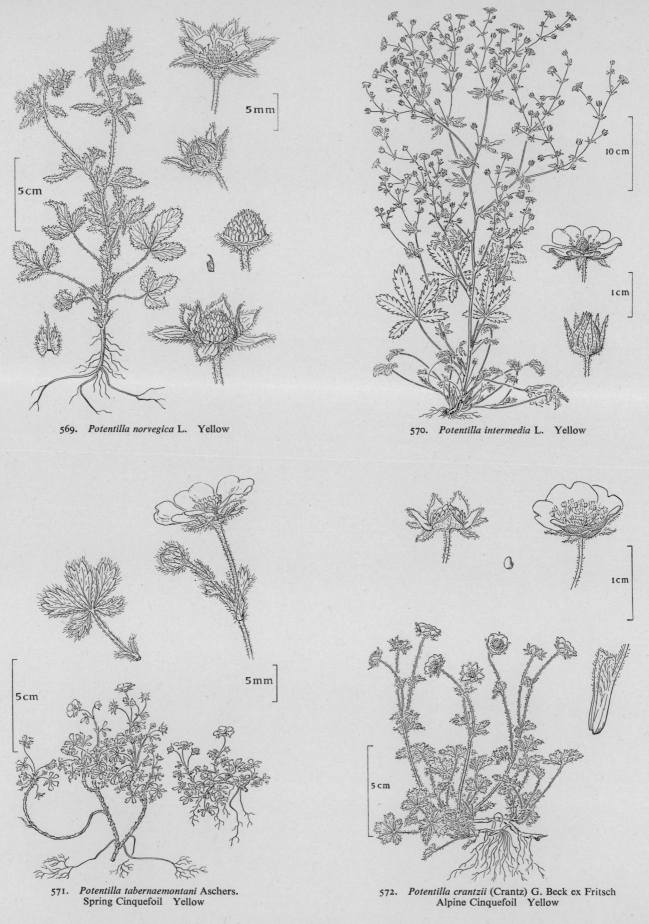

569. *Potentilla norvegica* L.   Yellow

570. *Potentilla intermedia* L.   Yellow

571. *Potentilla tabernaemontani* Aschers.
Spring Cinquefoil   Yellow

572. *Potentilla crantzii* (Crantz) G. Beck ex Fritsch
Alpine Cinquefoil   Yellow

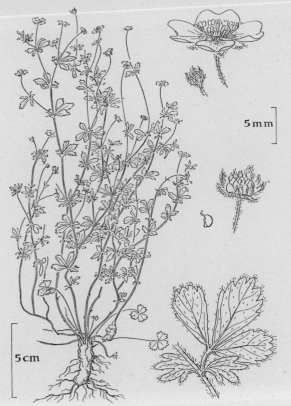

573. *Potentilla erecta* (L.) Räusch.   Common Tormentil
Yellow

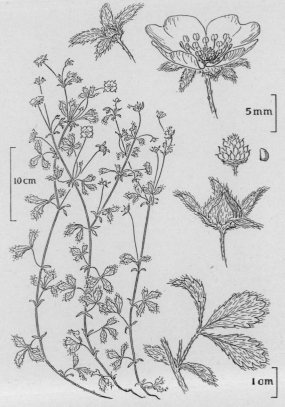

574. *Potentilla anglica* Laicharding   Trailing Tormentil
Yellow

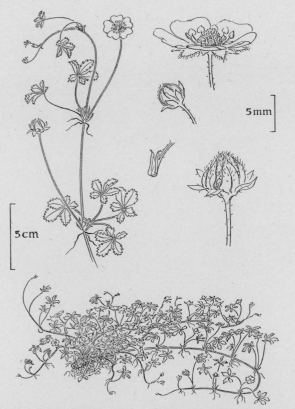

575. *Potentilla reptans* L.   Creeping Cinquefoil   Yellow

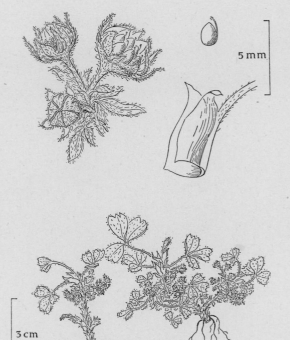

576. *Sibbaldia procumbens* L.   Greenish

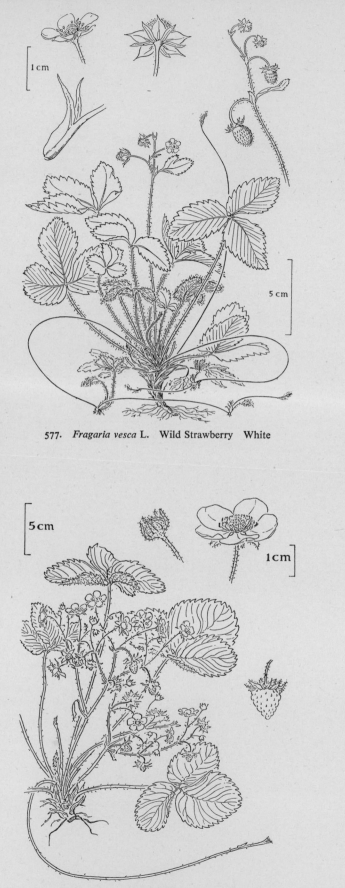

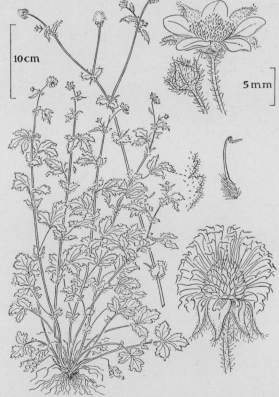

577. *Fragaria vesca* L. Wild Strawberry White

578. *Fragaria moschata* Duchesne
Hautbois Strawberry White

579. *Fragaria ananassa* Duchesne Garden Strawberry
White

580. *Geum urbanum* L. Herb Bennet Yellow

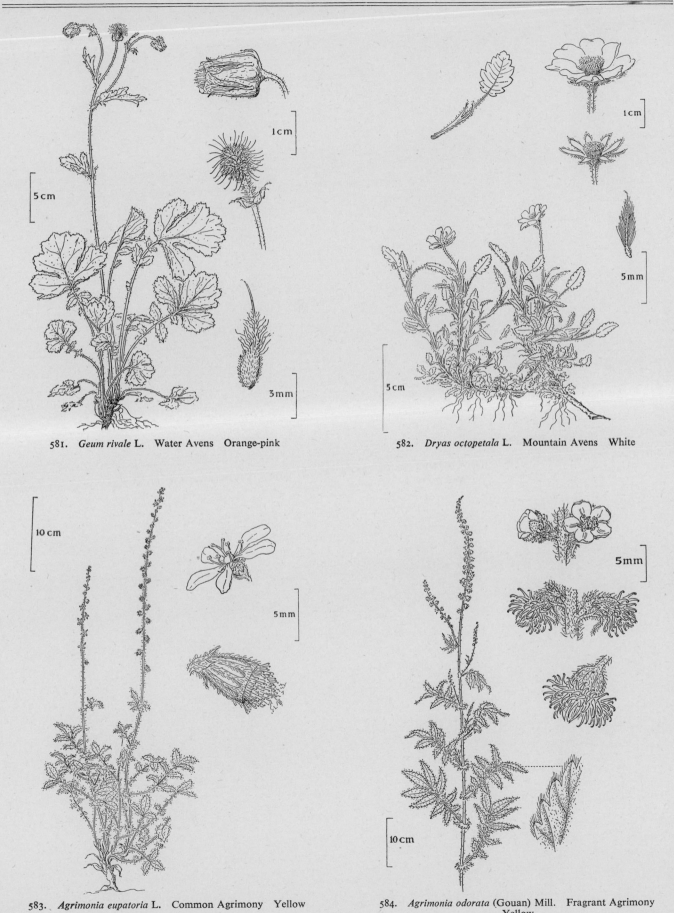

581. *Geum rivale* L. Water Avens Orange-pink

582. *Dryas octopetala* L. Mountain Avens White

583. *Agrimonia eupatoria* L. Common Agrimony Yellow

584. *Agrimonia odorata* (Gouan) Mill. Fragrant Agrimony
      Yellow

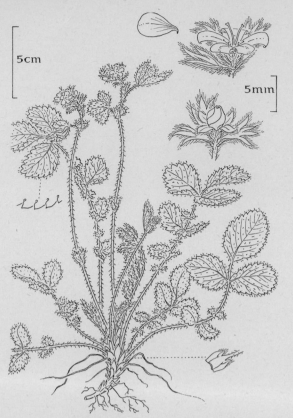

585. *Aremonia agrimonoides* (L.) DC.   Yellow

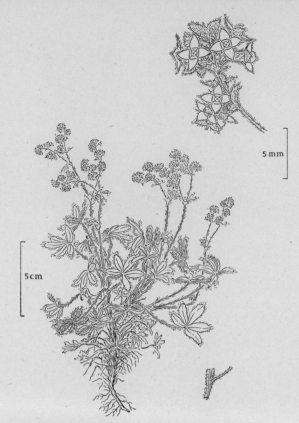

586. *Alchemilla alpina* L.   Alpine Lady's-Mantle   Greenish

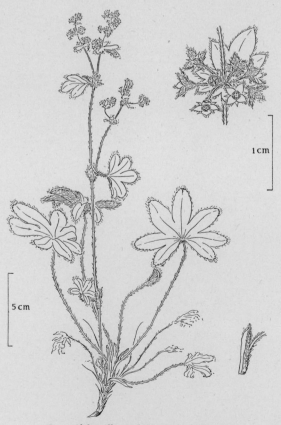

587. *Alchemilla conjuncta* Bab.   Greenish

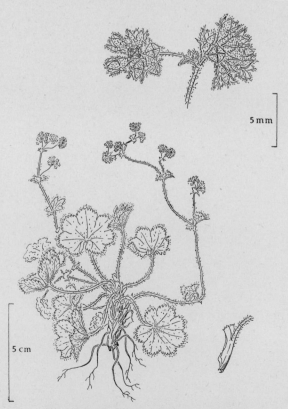

588. *Alchemilla glaucescens* Wallr.   Greenish

589. *Alchemilla vestita* (Buser) Raunk. Greenish

590. *Alchemilla filicaulis* Buser Greenish

591. *Alchemilla subcrenata* Buser Greenish

592. *Alchemilla minima* Walters Greenish

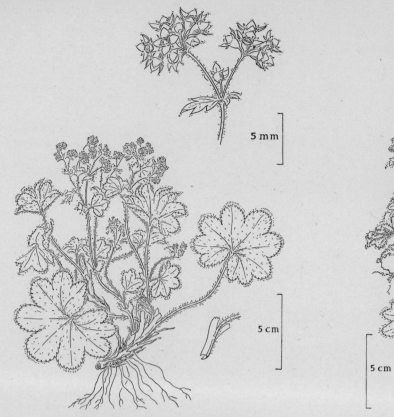

593. *Alchemilla monticola* Opiz   Greenish

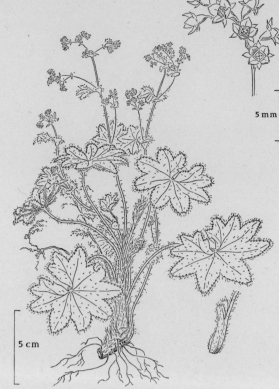

594. *Alchemilla acutiloba* Opiz   Greenish

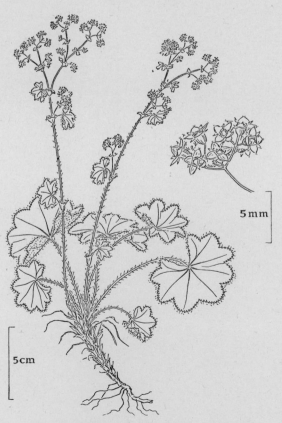

595. *Alchemilla xanthochlora* Rothm.   Greenish

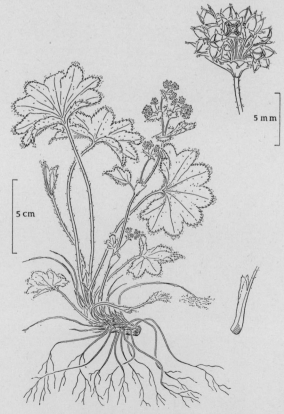

596. *Alchemilla glomerulans* Buser   Greenish

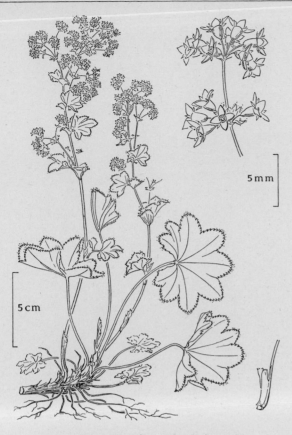

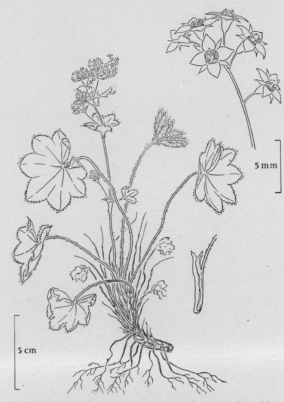

597. *Alchemilla glabra* Neygenf.   Greenish

598. *Alchemilla wichurae* (Buser) Stefánsson   Greenish

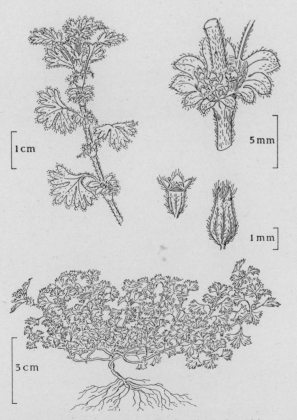

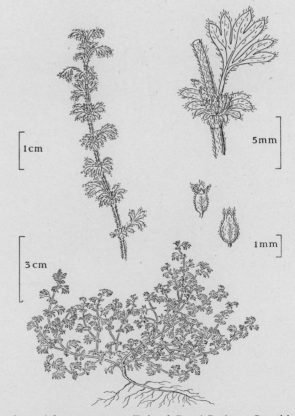

599. *Aphanes arvensis* L. Parsley Piert Greenish

600. *Aphanes microcarpa* (Boiss. & Reut.) Rothm.   Greenish

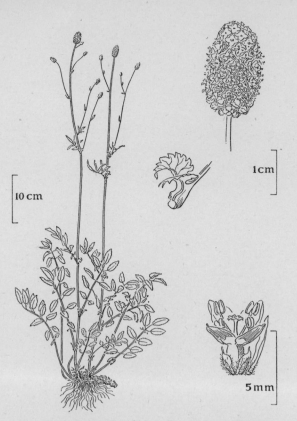

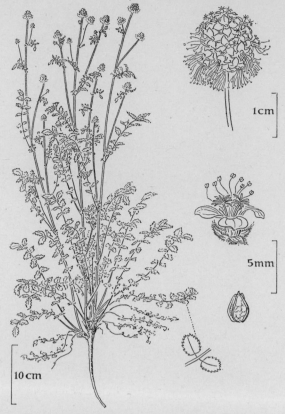

601.   *Sanguisorba officinalis* L.   Great Burnet   Crimson

602.   *Poterium sanguisorba* L.   Salad Burnet   Greenish

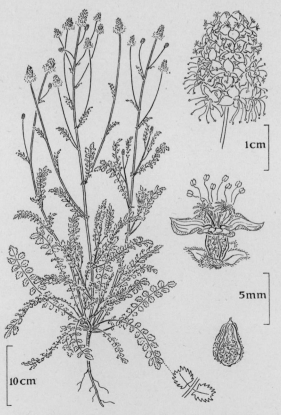

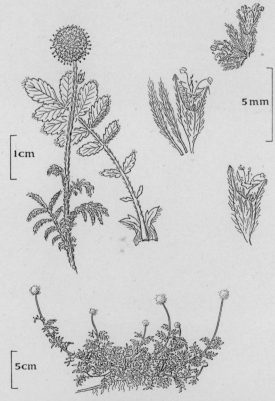

603.   *Poterium polygamum* Waldst. & Kit.   Greenish

604.   *Acaena anserinifolia* (J. R. & G. Forst.) Druce
Greenish

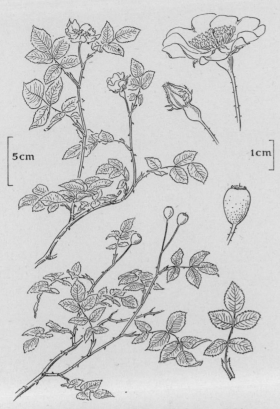

605. *Rosa arvensis* Huds. Field Rose White

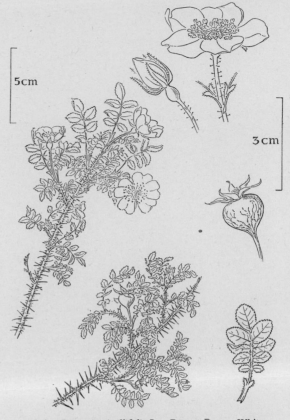

606. *Rosa pimpinellifolia* L. Burnet Rose White

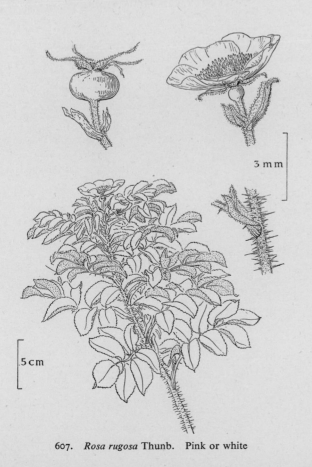

607. *Rosa rugosa* Thunb. Pink or white

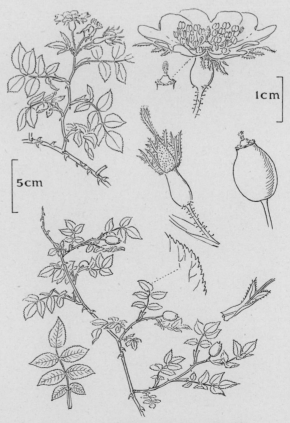

608. *Rosa stylosa* Desv. White or pale pink

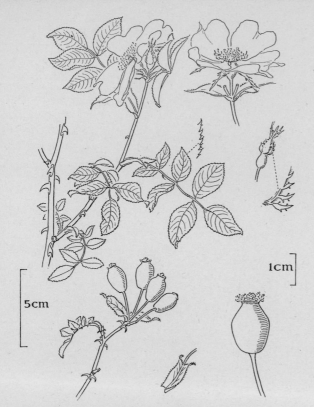

609. *Rosa canina* L. Dog Rose Pink or white

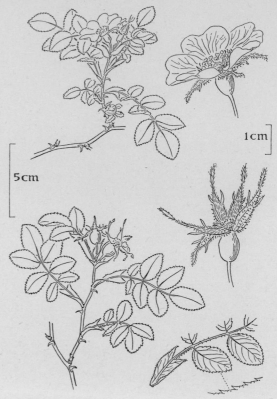

610. *Rosa dumalis* Bechst. Pink or white

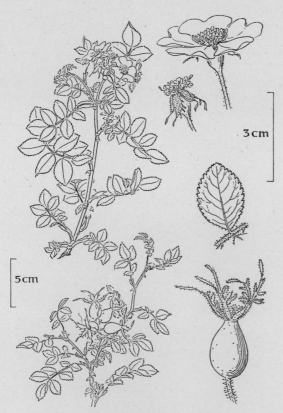

611. *Rosa obtusifolia* Desv. White

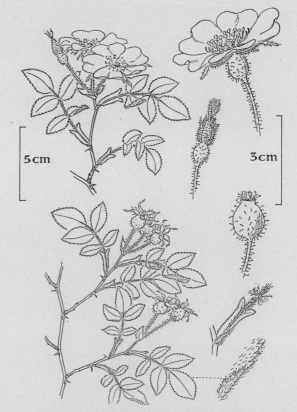

612. *Rosa tomentosa* Sm. Pink or white

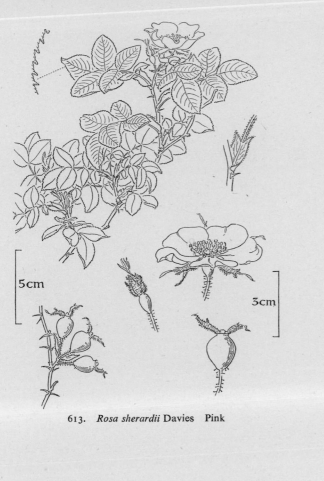

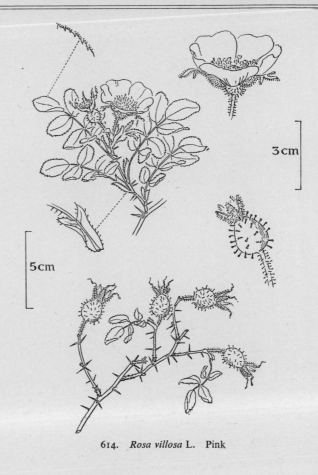

613. *Rosa sherardii* Davies   Pink

614. *Rosa villosa* L.   Pink

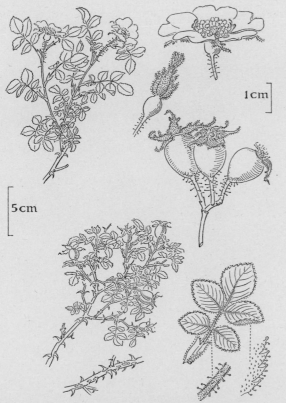

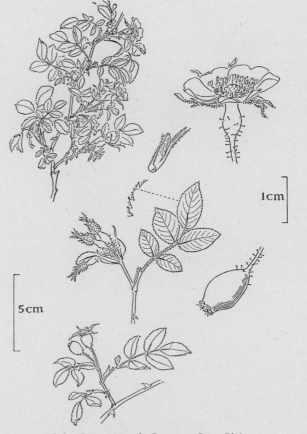

615. *Rosa rubiginosa* L.   Sweet-briar   Pink

616. *Rosa micrantha* Borrer ex Sm.   Pink

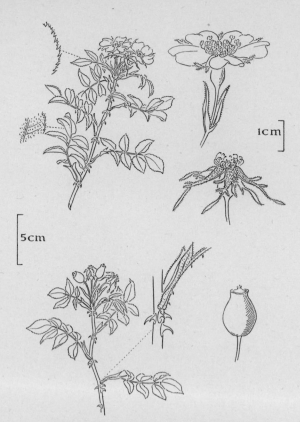

617. *Rosa agrestis* Savi   White or pale pink

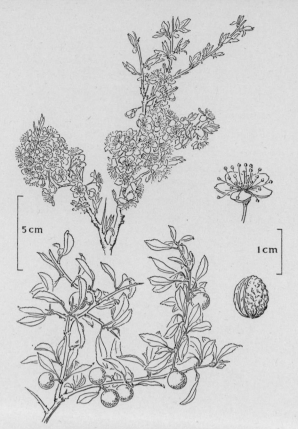

618. *Prunus spinosa* L.   Blackthorn, Sloe   White

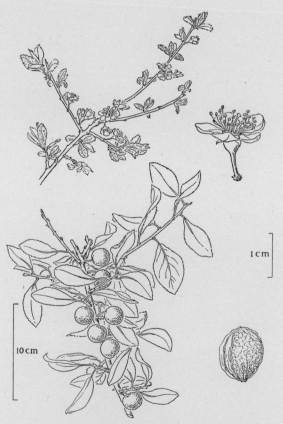

619. *Prunus domestica* L. ssp. *insititia* (L.) C. K. Schneid.
Bullace   White

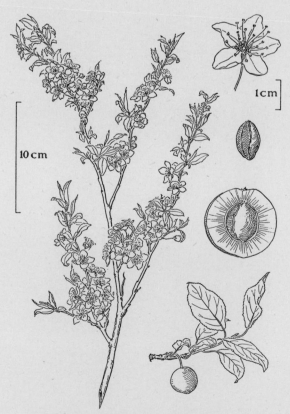

620. *Prunus cerasifera* Ehrh.   Cherry-plum   White

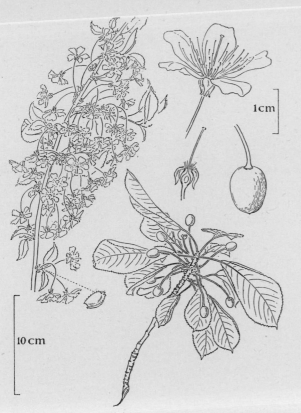

621. *Prunus avium* (L.) L.   Gean, Wild Cherry   White

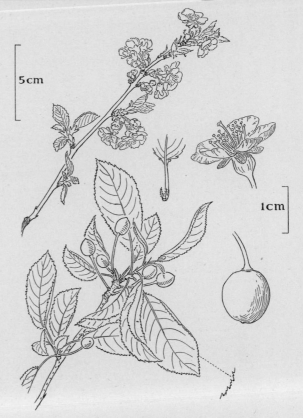

622. *Prunus cerasus* L.   Sour Cherry   White

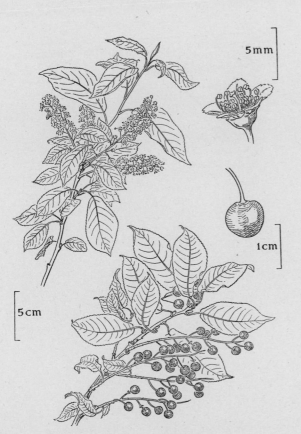

623. *Prunus padus* L.   Bird-Cherry   White

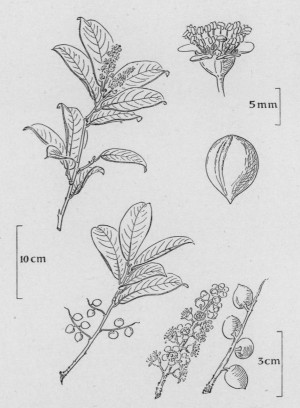

624. *Prunus laurocerasus* L.   Cherry-Laurel   White

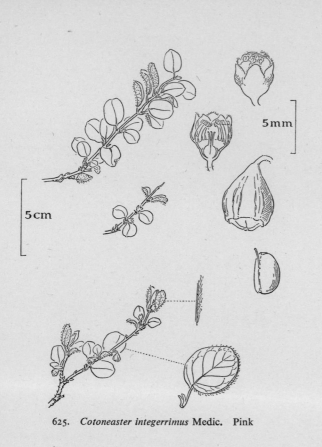

625.   *Cotoneaster integerrimus* Medic.   Pink

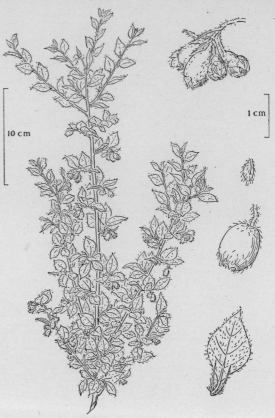

626.   *Cotoneaster simonsii* Baker   Pink

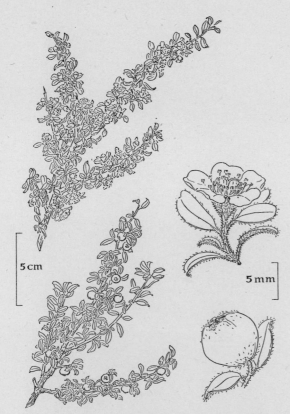

627.   *Cotoneaster microphyllus* Wall. ex Lindl.   White

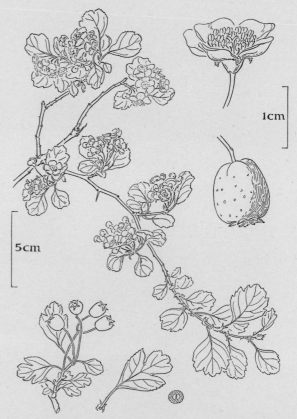

628.   *Crataegus oxyacanthoides* Thuill.   Midland Hawthorn
White

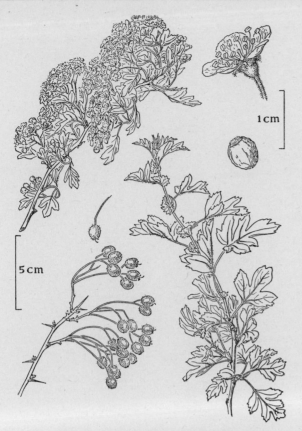

629. *Crataegus monogyna* Jacq.   Hawthorn   White

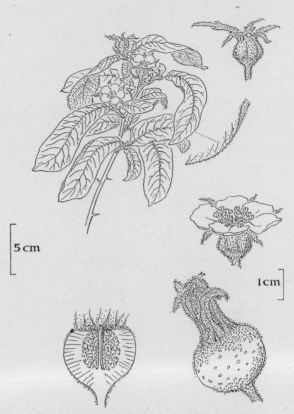

630. *Mespilus germanica* L.   Medlar   White

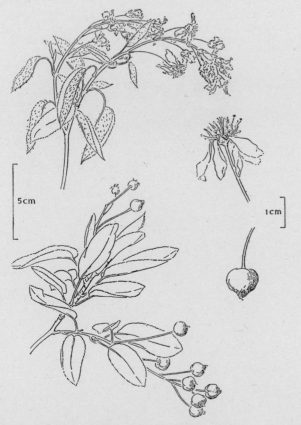

631. *Amelanchier laevis* Wieg.   White

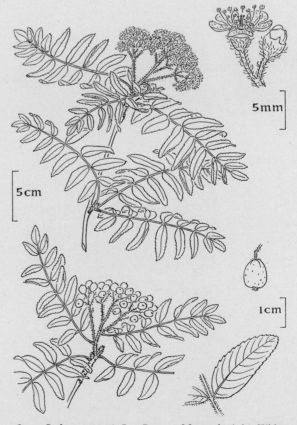

632 *Sorbus aucuparia* L.   Rowan, Mountain Ash   White

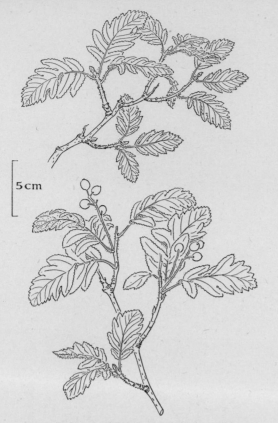

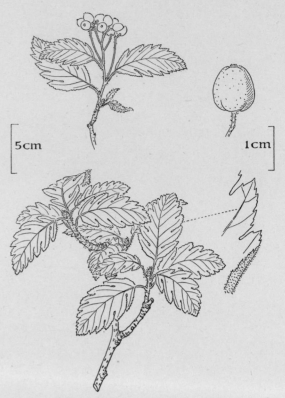

633. *Sorbus pseudofennica* E. F. Warb. White

634. *Sorbus arranensis* Hedl. White

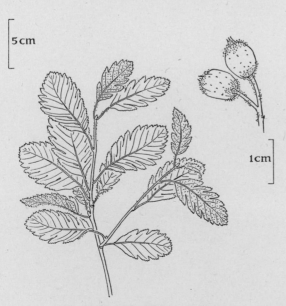

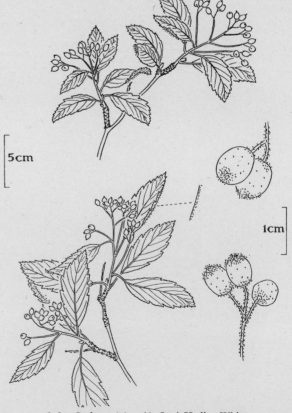

635. *Sorbus leyana* Wilmott White

636. *Sorbus minima* (A. Ley) Hedl. White

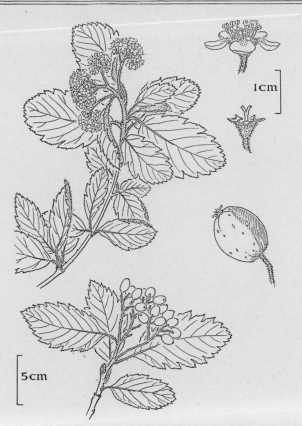

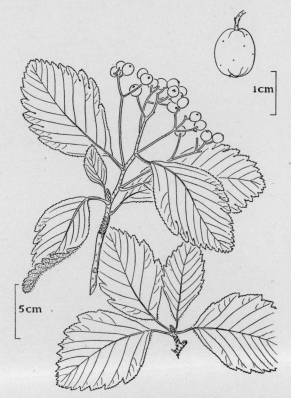

637.   *Sorbus intermedia* (Ehrh.) Pers.   White

638.   *Sorbus anglica* Hedl.   White

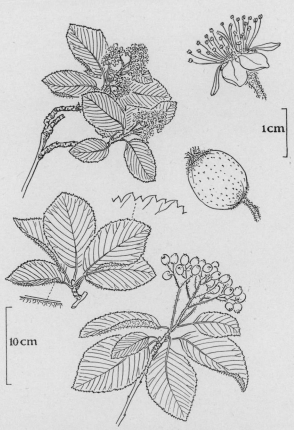

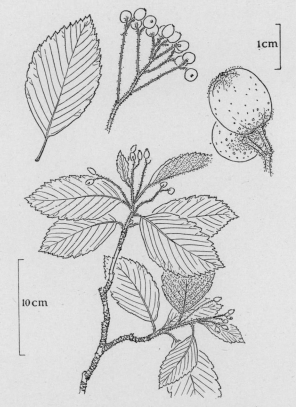

639.   *Sorbus aria* (L.) Crantz   White Beam   White

640.   *Sorbus leptophylla* E. F. Warb.   White

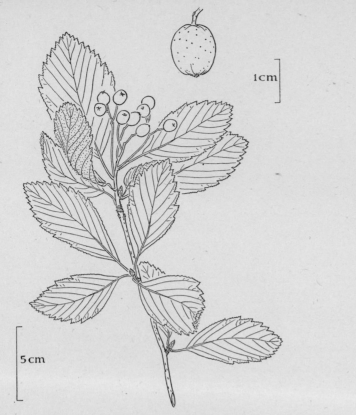

1cm

5cm

641.   *Sorbus wilmottiana* E. F. Warb.   White

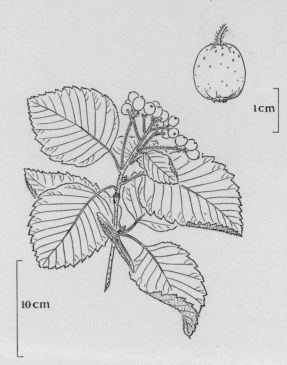

1cm

10cm

642.   *Sorbus eminens* E. F. Warb.   White

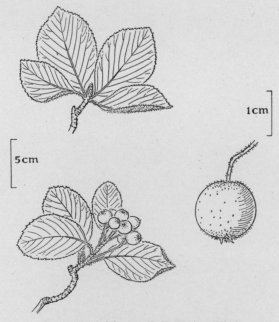

1cm

5cm

643.   *Sorbus hibernica* E. F. Warb.   White

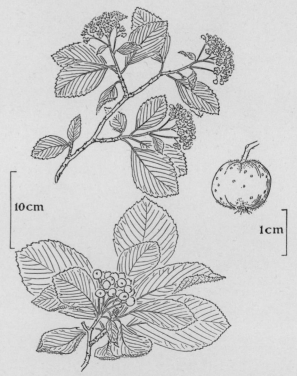

10cm

1cm

644.   *Sorbus porrigentiformis* E. F. Warb.   White

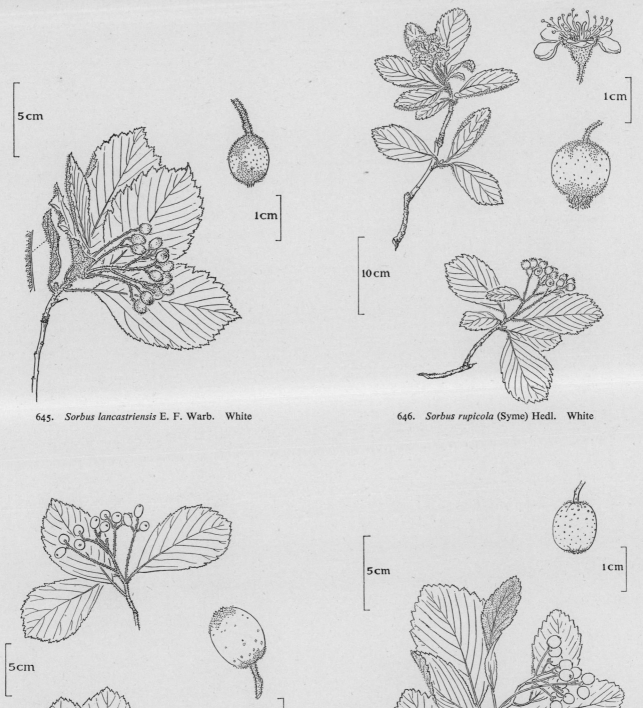

645. *Sorbus lancastriensis* E. F. Warb.   White

646. *Sorbus rupicola* (Syme) Hedl.   White

647. *Sorbus vexans* E. F. Warb.   White

648. *Sorbus bristoliensis* Wilmott   White

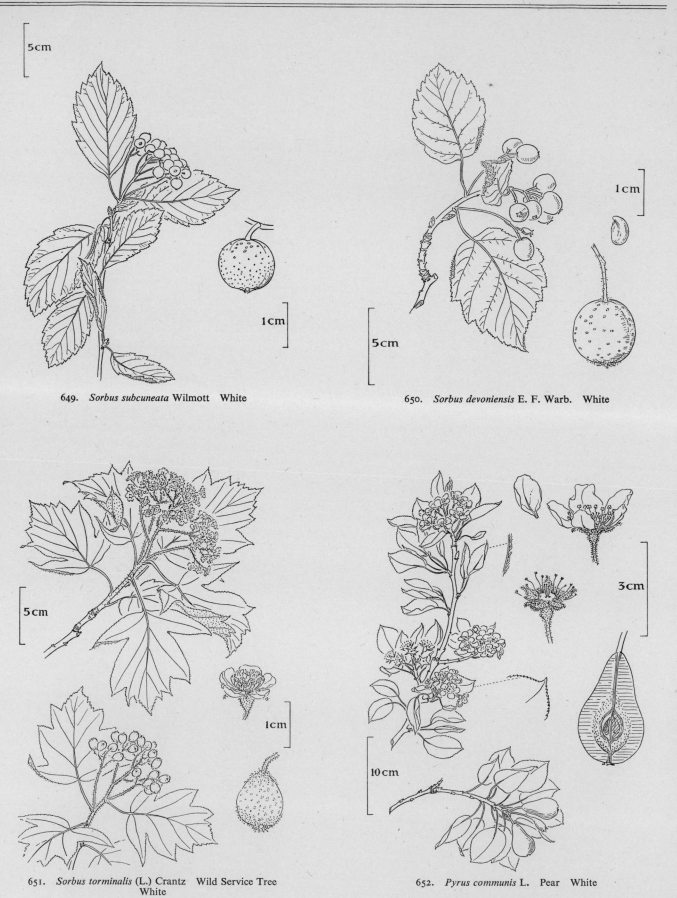

649. *Sorbus subcuneata* Wilmott   White

650. *Sorbus devoniensis* E. F. Warb.   White

651. *Sorbus torminalis* (L.) Crantz   Wild Service Tree
White

652. *Pyrus communis* L.   Pear   White

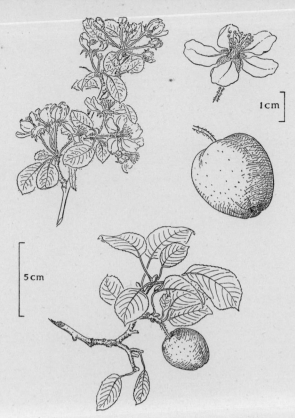

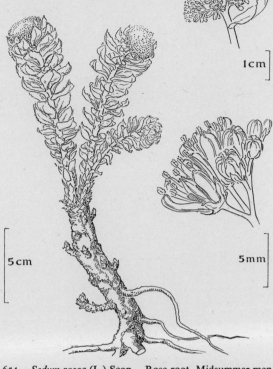

653. *Malus sylvestris* Mill. Crab Apple Pinkish

654. *Sedum rosea* (L.) Scop. Rose-root, Midsummer-men
Greenish

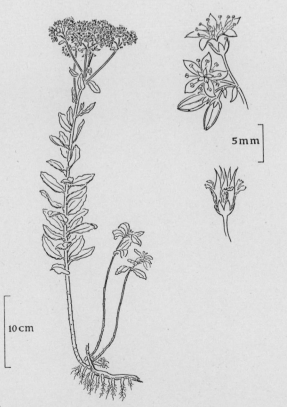

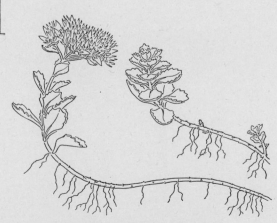

655. *Sedum telephium* L. Orpine, Livelong
Reddish-purple

656. *Sedum spurium* M. Bieb. Pink

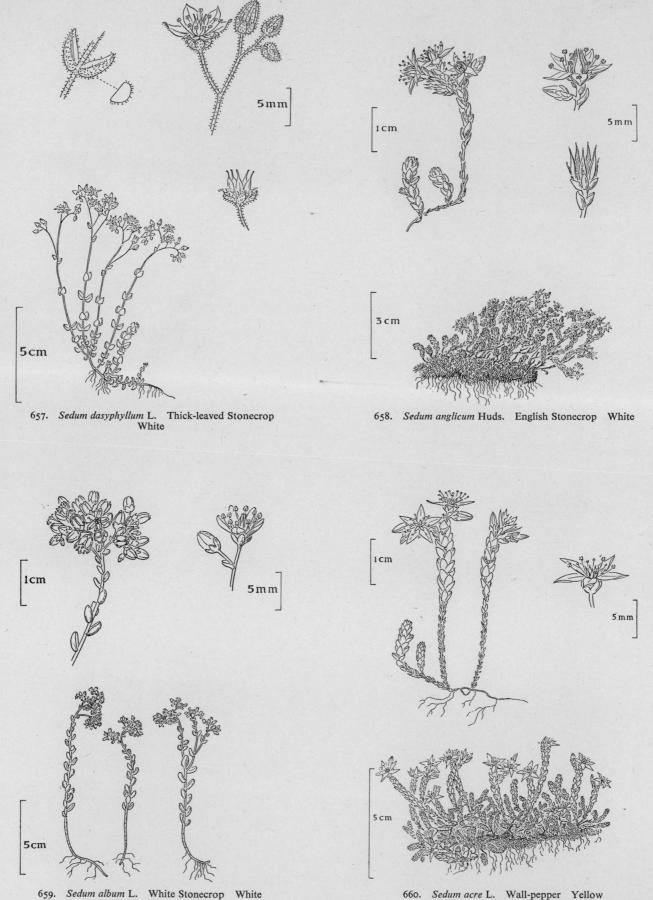

657. *Sedum dasyphyllum* L.   Thick-leaved Stonecrop
White

658. *Sedum anglicum* Huds.   English Stonecrop   White

659. *Sedum album* L.   White Stonecrop   White

660. *Sedum acre* L.   Wall-pepper   Yellow

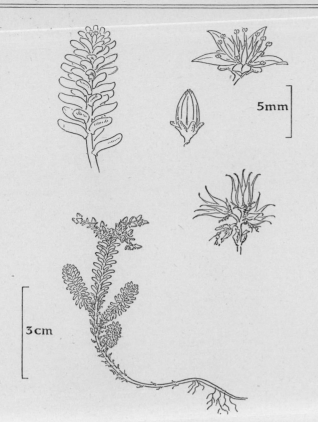

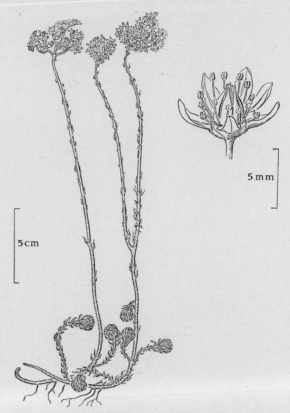

661. *Sedum sexangulare* L.   Insipid Stonecrop   Yellow

662. *Sedum forsteranum* Sm.   Rock Stonecrop   Yellow

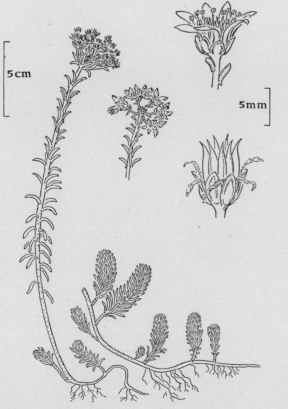

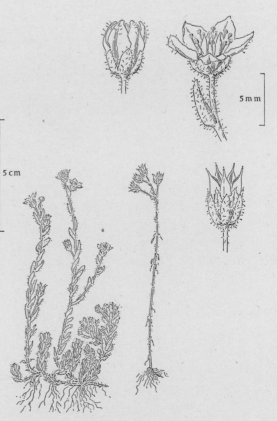

663. *Sedum reflexum* L.   Yellow

664. *Sedum villosum* L.   Hairy Stonecrop   Pink

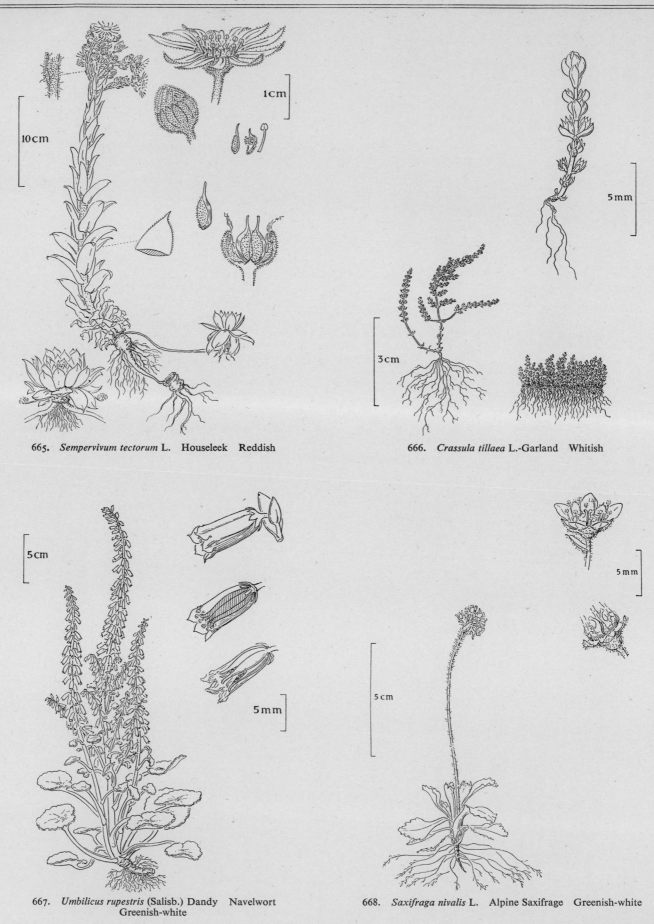

665.   *Sempervivum tectorum* L.   Houseleek   Reddish

666.   *Crassula tillaea* L.-Garland   Whitish

667.   *Umbilicus rupestris* (Salisb.) Dandy   Navelwort
Greenish-white

668.   *Saxifraga nivalis* L.   Alpine Saxifrage   Greenish-white

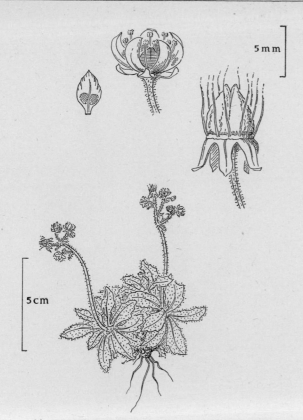

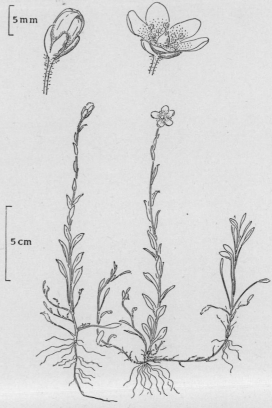

669. *Saxifraga stellaris* L.   Starry Saxifrage   White

670. *Saxifraga hirculus* L.   Yellow Marsh Saxifrage   Yellow

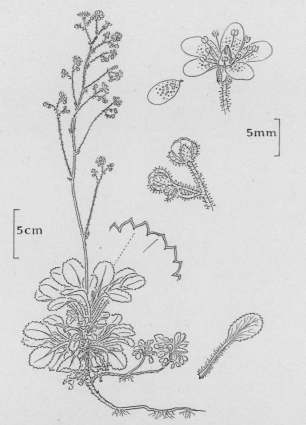

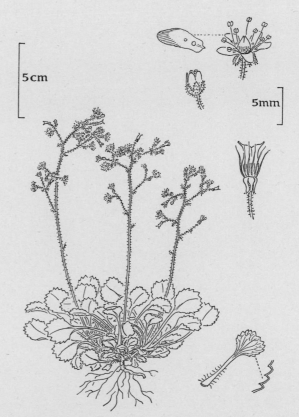

671. *Saxifraga spathularis × umbrosa*   London Pride   White

672. *Saxifraga spathularis* Brot.   St Patrick's Cabbage   White

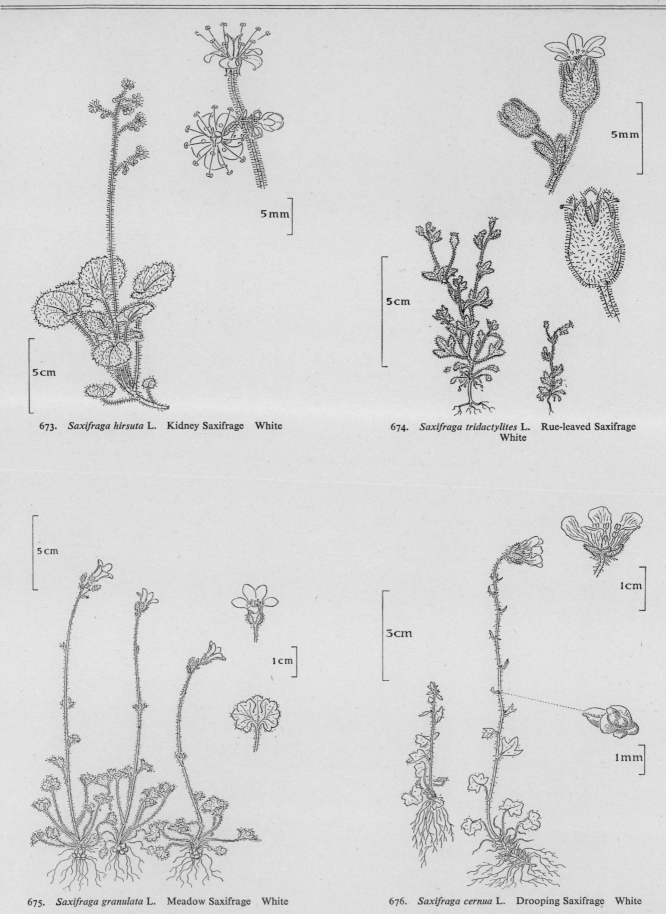

673.   *Saxifraga hirsuta* L.   Kidney Saxifrage   White

674.   *Saxifraga tridactylites* L.   Rue-leaved Saxifrage
White

675.   *Saxifraga granulata* L.   Meadow Saxifrage   White

676.   *Saxifraga cernua* L.   Drooping Saxifrage   White

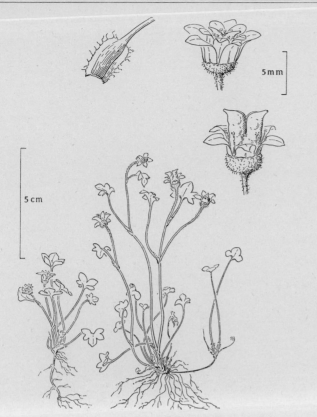

677. *Saxifraga rivularis* L.   Brook Saxifrage   White

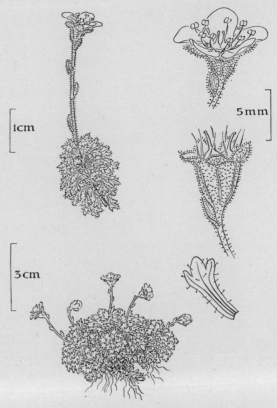

678. *Saxifraga cespitosa* L.   Tufted Saxifrage   White

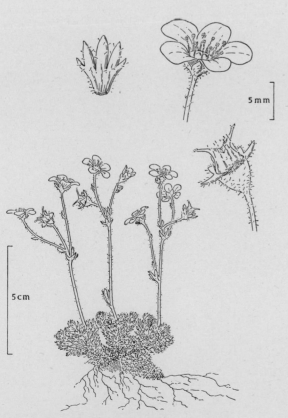

679. *Saxifraga rosacea* Moench   White

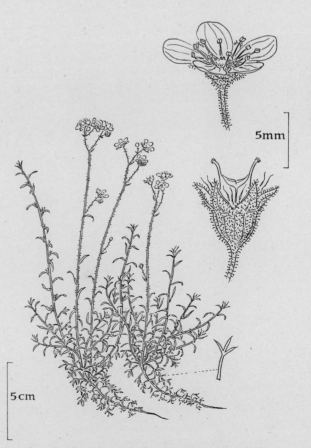

680. *Saxifraga hypnoides* L.   Dovedale Moss   White

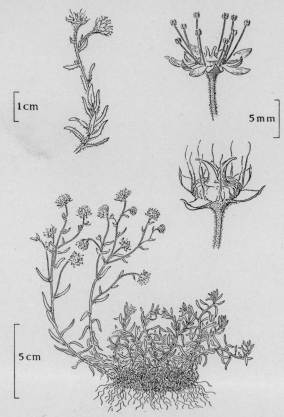

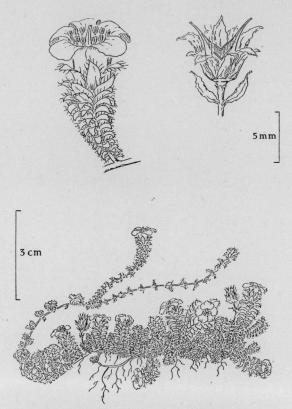

681. *Saxifraga aizoides* L.   Yellow Mountain Saxifrage
     Yellow

682. *Saxifraga oppositifolia* L.   Purple Saxifrage   Purple

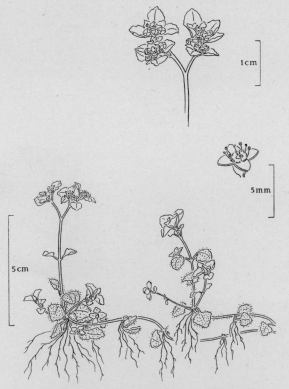

683. *Chrysosplenium oppositifolium* L.   Opposite-leaved
     Golden Saxifrage   Greenish-yellow

684. *Chrysosplenium alternifolium* L.   Alternate-leaved
     Golden Saxifrage   Greenish-yellow

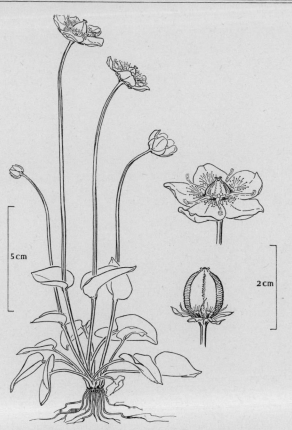

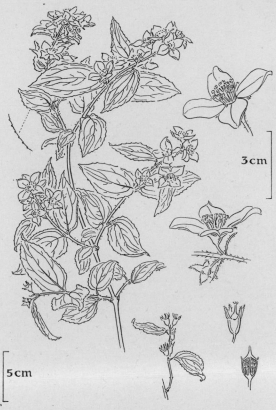

685. *Parnassia palustris* L. Grass of Parnassus White

686. *Philadelphus coronarius* L. Syringa, Mock Orange
White

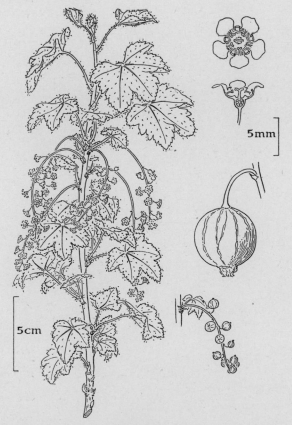

687. *Ribes sylvestre* (Lam.) Mert. & Koch Red Currant
Greenish

688. *Ribes spicatum* Robson Greenish

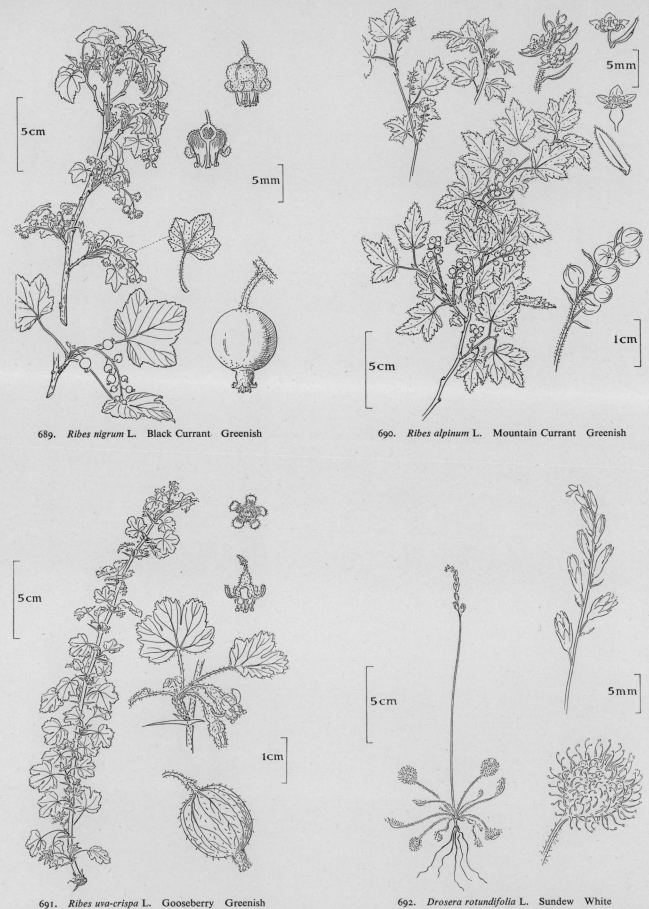

689.   *Ribes nigrum* L.   Black Currant   Greenish

690.   *Ribes alpinum* L.   Mountain Currant   Greenish

691.   *Ribes uva-crispa* L.   Gooseberry   Greenish

692.   *Drosera rotundifolia* L.   Sundew   White

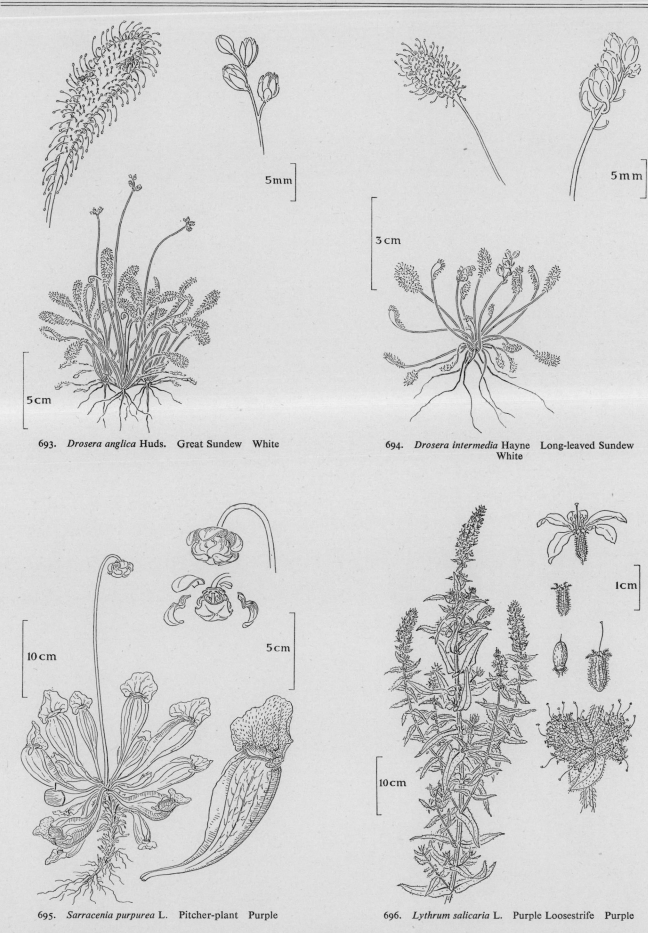

5 mm

5 cm

693.   *Drosera anglica* Huds.   Great Sundew   White

5 mm

3 cm

694.   *Drosera intermedia* Hayne   Long-leaved Sundew
White

10 cm

5 cm

695.   *Sarracenia purpurea* L.   Pitcher-plant   Purple

1 cm

10 cm

696.   *Lythrum salicaria* L.   Purple Loosestrife   Purple

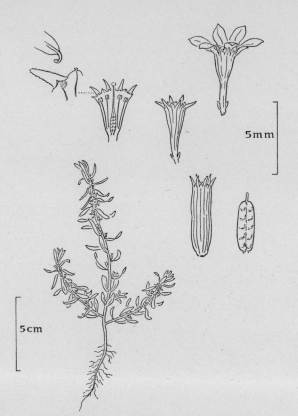

697. *Lythrum hyssopifolia* L.   Grass Poly   Pink

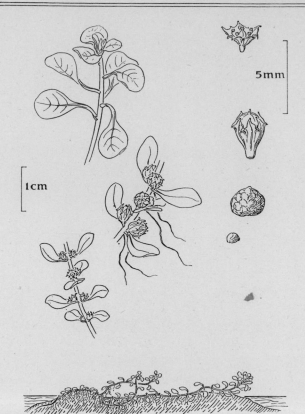

698. *Peplis portula* L.   Water Purslane   Greenish

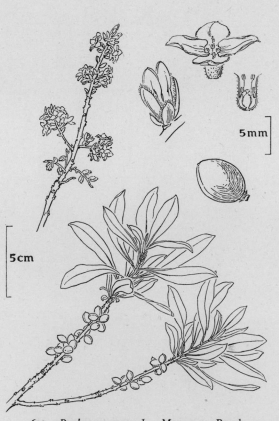

699. *Daphne mezereum* L.   Mezereon   Purple

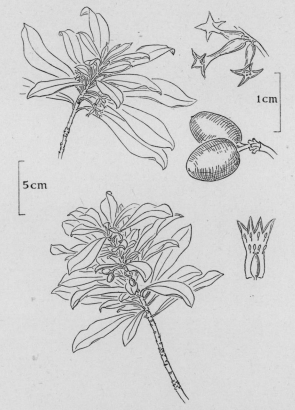

700. *Daphne laureola* L.   Spurge-laurel   Green

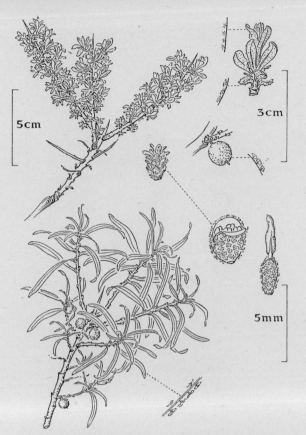

701. *Hippophaë rhamnoides* L.　Sea Buckthorn　Greenish

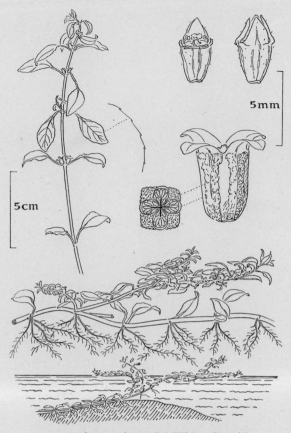

702. *Ludwigia palustris* (L.) Elliott　Greenish

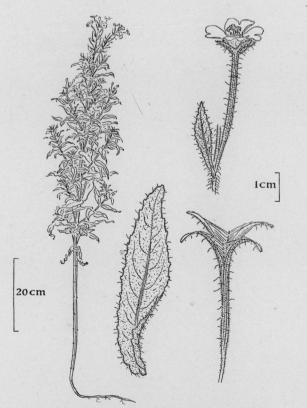

703. *Epilobium hirsutum* L.　Great Hairy Willow-herb,
Codlins and Cream　Purplish

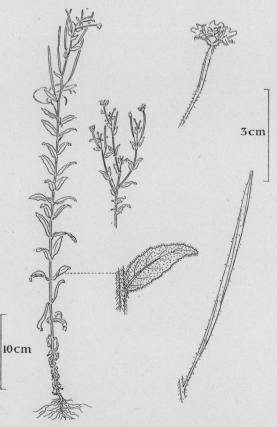

704. *Epilobium parviflorum* Schreb.　Small-flowered Hairy
Willow-herb　Purplish

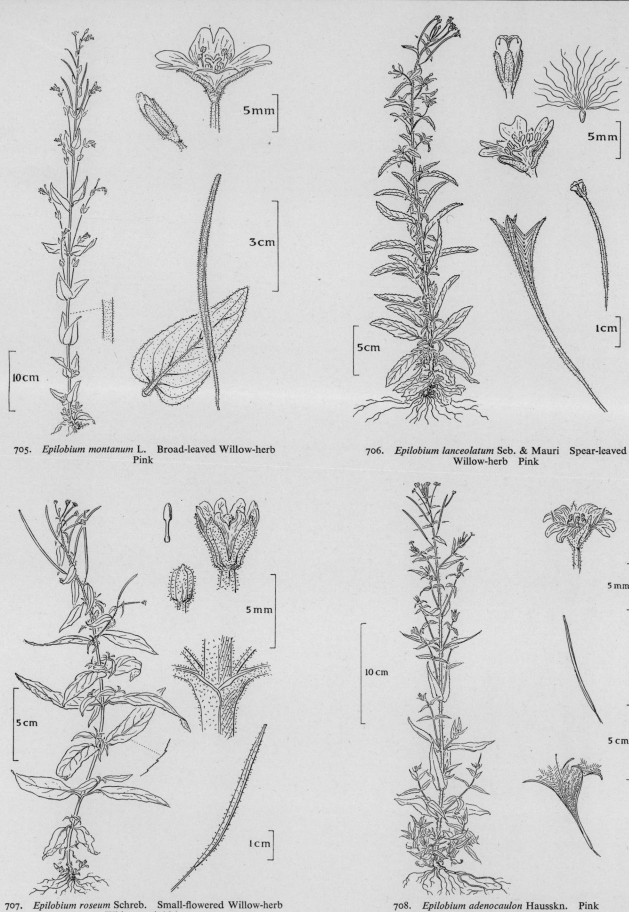

705. *Epilobium montanum* L.   Broad-leaved Willow-herb
      Pink

706. *Epilobium lanceolatum* Seb. & Mauri   Spear-leaved
      Willow-herb   Pink

707. *Epilobium roseum* Schreb.   Small-flowered Willow-herb
      White or pinkish

708. *Epilobium adenocaulon* Hausskn.   Pink

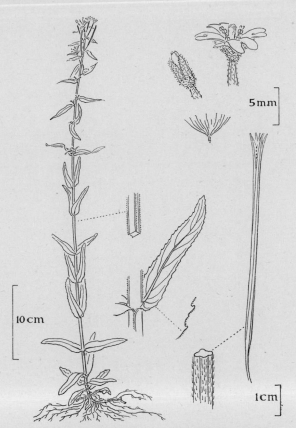

5mm

10cm

1cm

709. *Epilobium adnatum* Griseb.   Square-stemmed Willow-
herb   Lilac

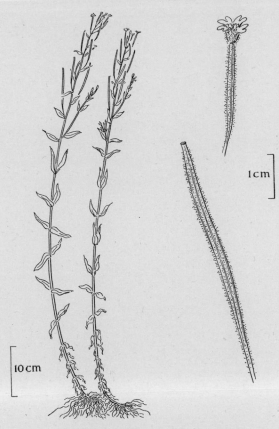

1cm

10cm

710. *Epilobium obscurum* Schreb.   Pink

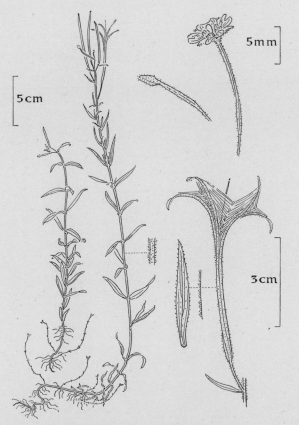

5mm

5cm

3cm

711. *Epilobium palustre* L.   Marsh Willow-herb
Pink or lilac

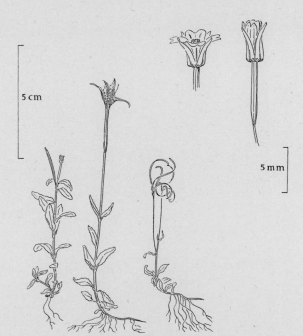

5cm

5mm

712. *Epilobium anagallidifolium* Lam.   Alpine Willow-herb
Pink

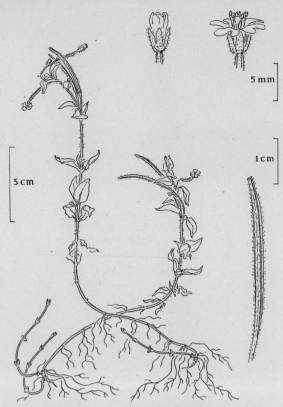

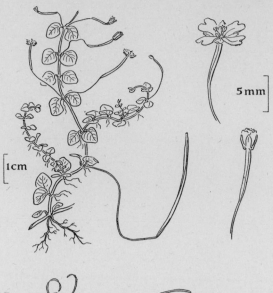

713.   *Epilobium alsinifolium* Vill.   Chickweed Willow-herb
       Reddish

714.   *Epilobium nerterioides* A. Cunn.   Pink

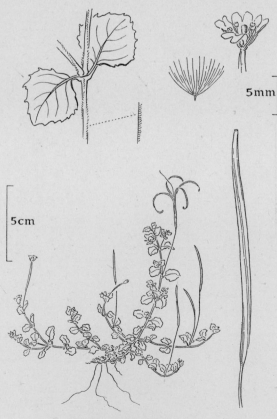

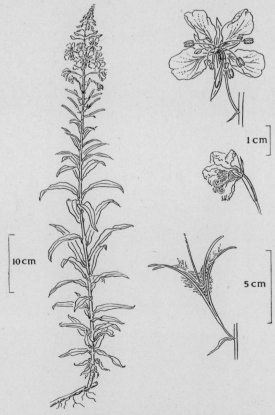

715.   *Epilobium pedunculare* A. Cunn.   Pink

716.   *Chamaenerion angustifolium* (L.) Scop.   Rosebay
       Willow-herb, Fireweed   Purple

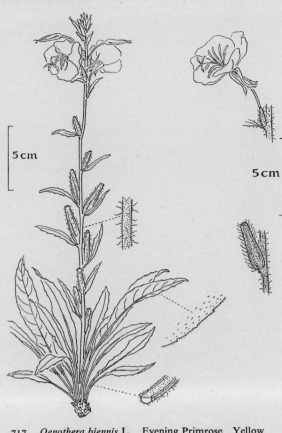

717.   *Oenothera biennis* L.   Evening Primrose   Yellow

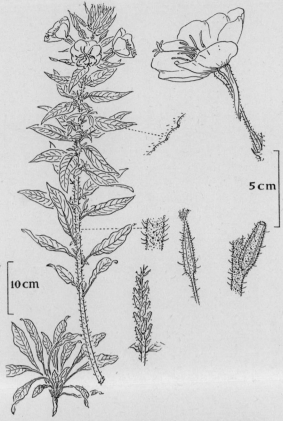

718.   *Oenothera erythrosepala* Borbás   Evening Primrose
Yellow

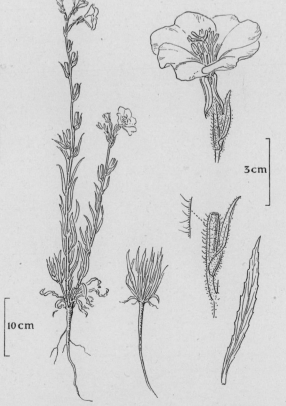

719.   *Oenothera stricta* Ledeb. ex Link   Evening Primrose
Yellow becoming red

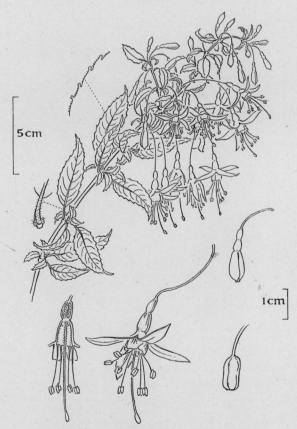

720.   *Fuchsia magellanica* Lam.   Willow Fuchsia
Red and violet

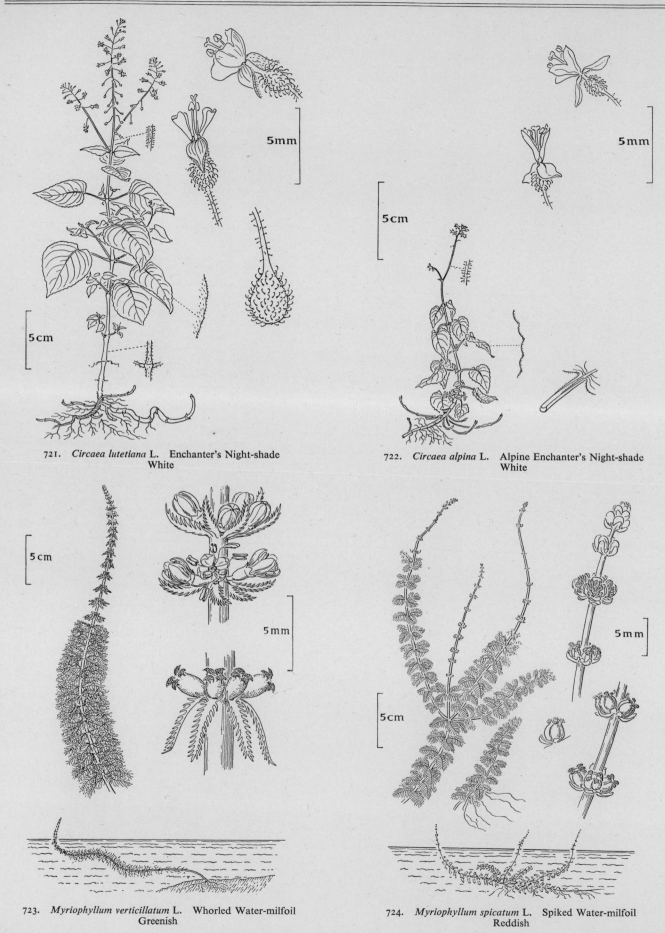

721. *Circaea lutetiana* L.   Enchanter's Night-shade
White

722. *Circaea alpina* L.   Alpine Enchanter's Night-shade
White

723. *Myriophyllum verticillatum* L.   Whorled Water-milfoil
Greenish

724. *Myriophyllum spicatum* L.   Spiked Water-milfoil
Reddish

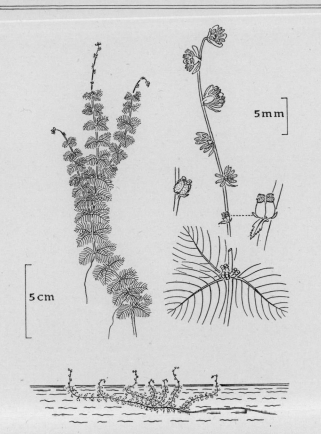

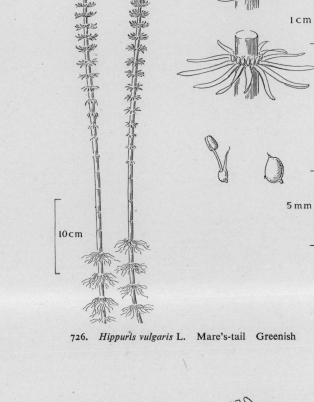

725. *Myriophyllum alterniflorum* DC.   Alternate-flowered
Water-milfoil   Yellowish

726. *Hippuris vulgaris* L.   Mare's-tail   Greenish

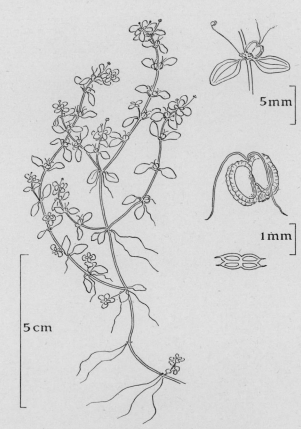

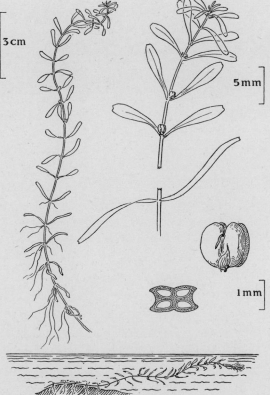

727. *Callitriche stagnalis* Scop.   Greenish

728. *Callitriche platycarpa* Kütz.   Greenish

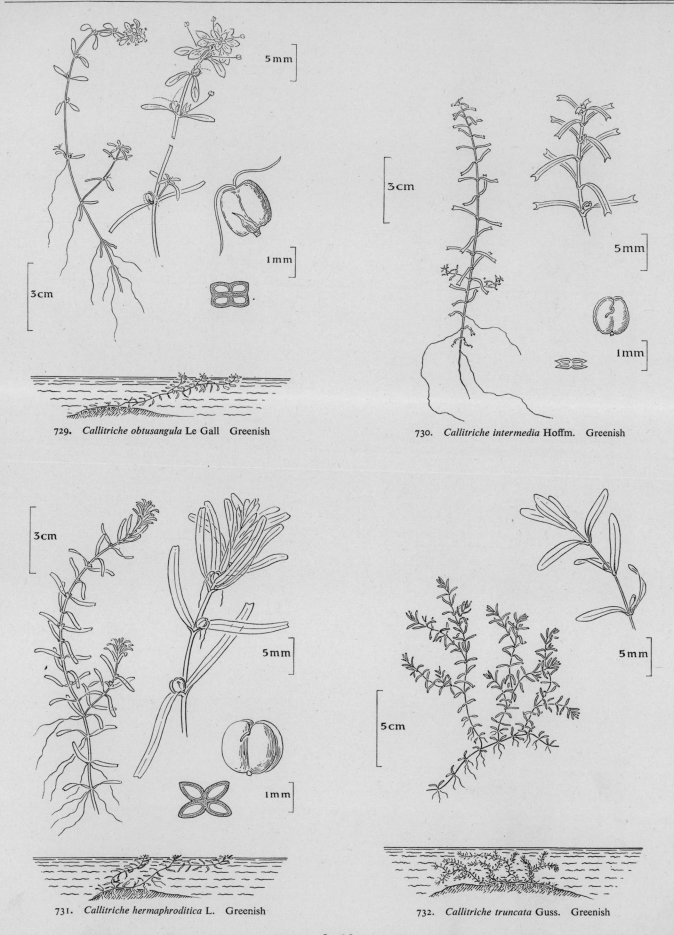

729. *Callitriche obtusangula* Le Gall   Greenish

730. *Callitriche intermedia* Hoffm.   Greenish

731. *Callitriche hermaphroditica* L.   Greenish

732. *Callitriche truncata* Guss.   Greenish

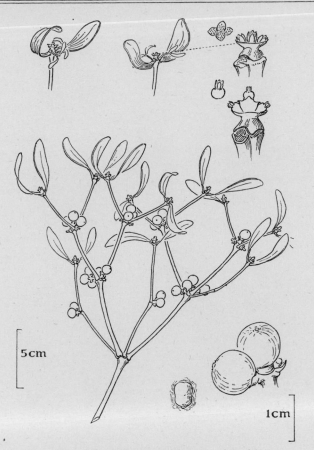

733. *Viscum album* L.   Mistletoe   Greenish

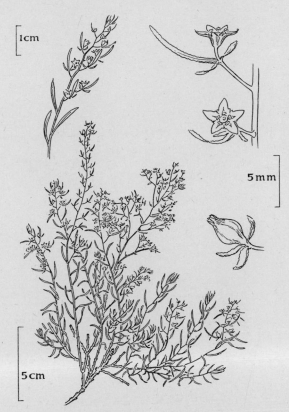

734. *Thesium humifusum* DC.   Bastard Toadflax   Yellowish

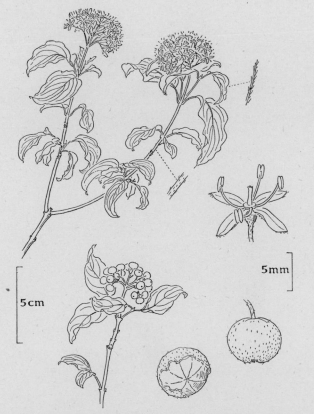

735. *Thelycrania sanguinea* (L.) Fourr.   Dogwood   White

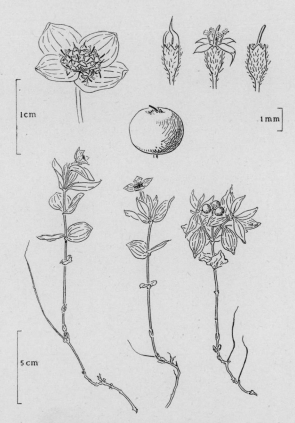

736. *Chamaepericlymenum suecicum* (L.) Aschers. & Graebn.
Dwarf Cornel   White and dark purple

737. *Hedera helix* L.   Ivy   Yellowish

738. *Hydrocotyle vulgaris* L.   Pennywort, White-rot
Greenish

739. *Sanicula europaea* L.   Sanicle   Pink or white

740. *Astrantia major* L.   Pink or white

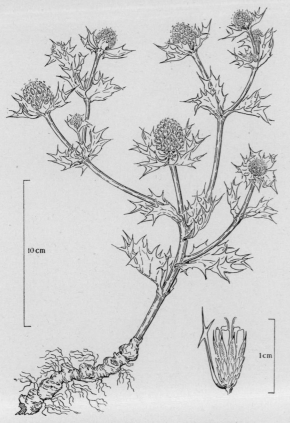

741. *Eryngium maritimum* L.   Sea Holly   Blue

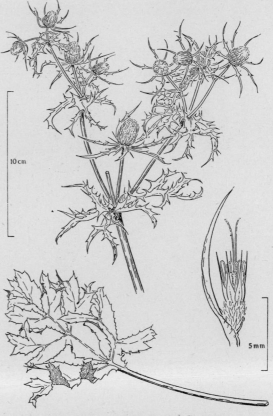

742. *Eryngium campestre* L.   Greenish

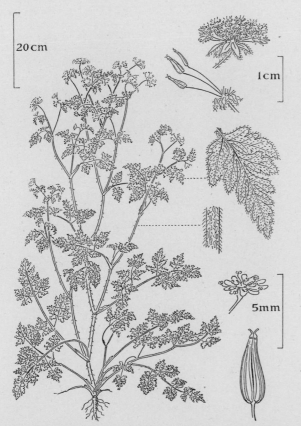

743. *Chaerophyllum temulentum* L.   Rough Chervil   White

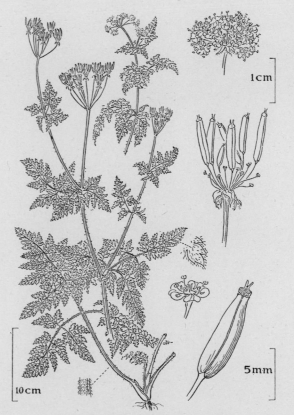

744. *Chaerophyllum aureum* L.   White

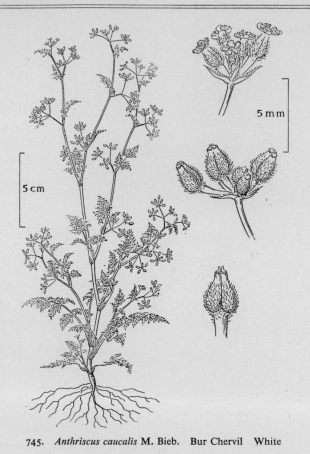

745. *Anthriscus caucalis* M. Bieb.   Bur Chervil   White

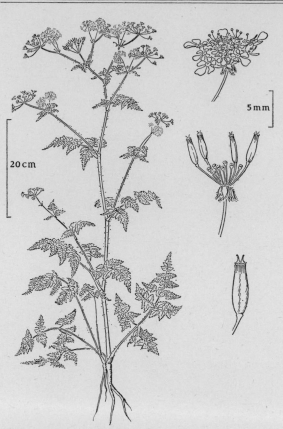

746. *Anthriscus sylvestris* (L.) Hoffm.   Cow Parsley, Keck
White

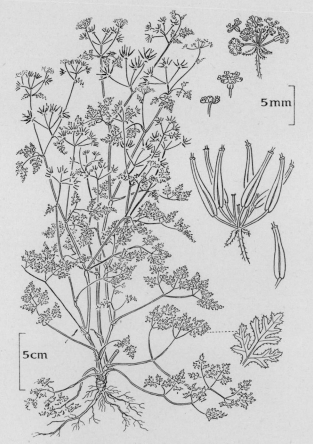

747. *Anthriscus cerefolium* (L.) Hoffm.   White

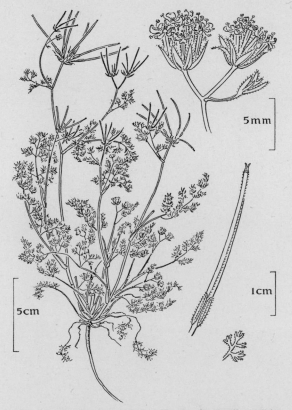

748. *Scandix pecten-veneris* L.   Shepherd's Needle   White

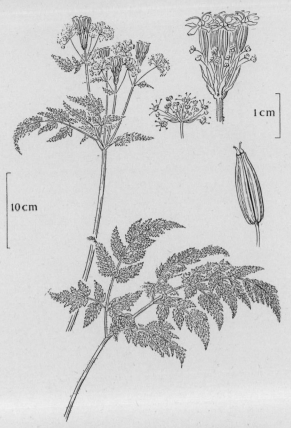

749. *Myrrhis odorata* (L.) Scop. Sweet Cicely White

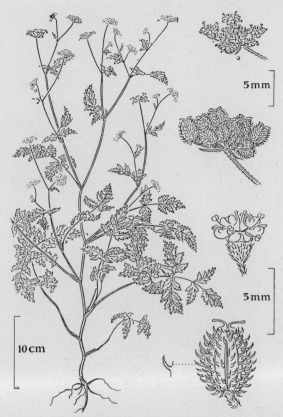

750. *Torilis japonica* (Houtt.) DC. Upright Hedge-parsley
White

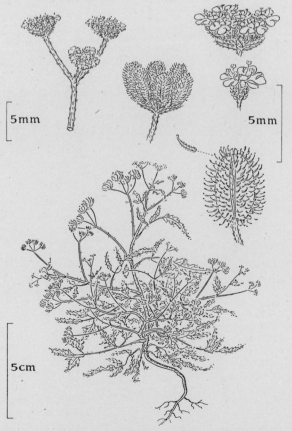

751. *Torilis arvensis* (Huds.) Link Spreading Hedge-parsley
White

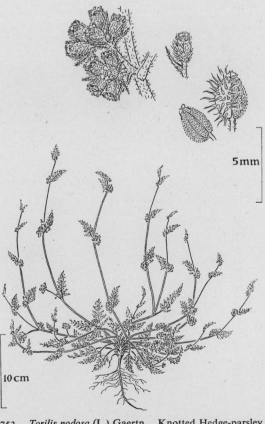

752. *Torilis nodosa* (L.) Gaertn. Knotted Hedge-parsley
White

4-2

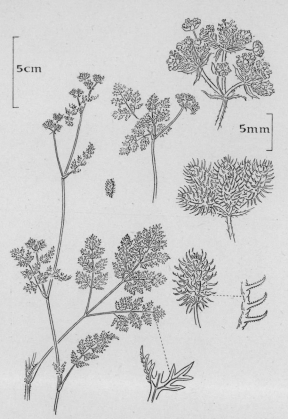

753. *Caucalis platycarpos* L.   Small Bur-parsley   White

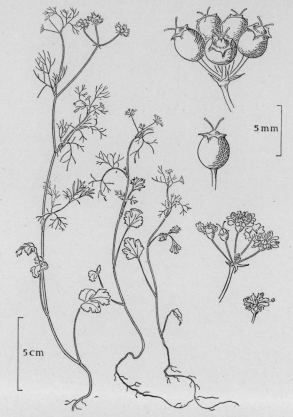

754. *Coriandrum sativum* L.   Coriander   White

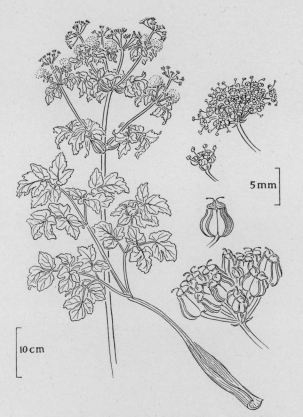

755. *Smyrnium olusatrum* L.   Alexanders   Yellowish

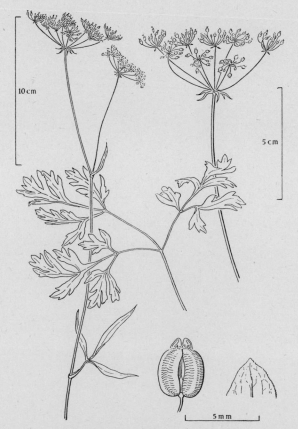

756. *Physospermum cornubiense* (L.) DC.   Bladder-seed
White

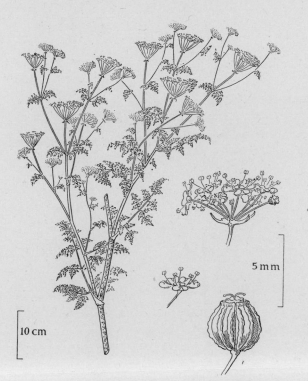

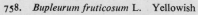

757. *Conium maculatum* L.   Hemlock   White

758. *Bupleurum fruticosum* L.   Yellowish

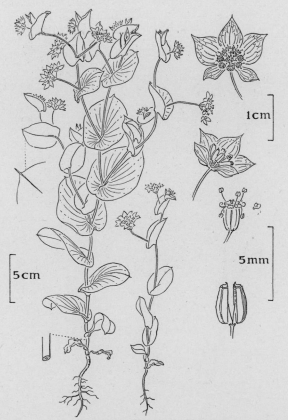

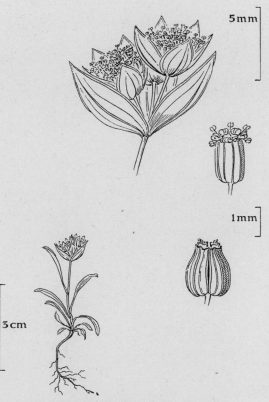

759. *Bupleurum rotundifolium* L.   Hare's-ear, Thorow-wax
Yellowish

760. *Bupleurum baldense* Turra   Yellowish

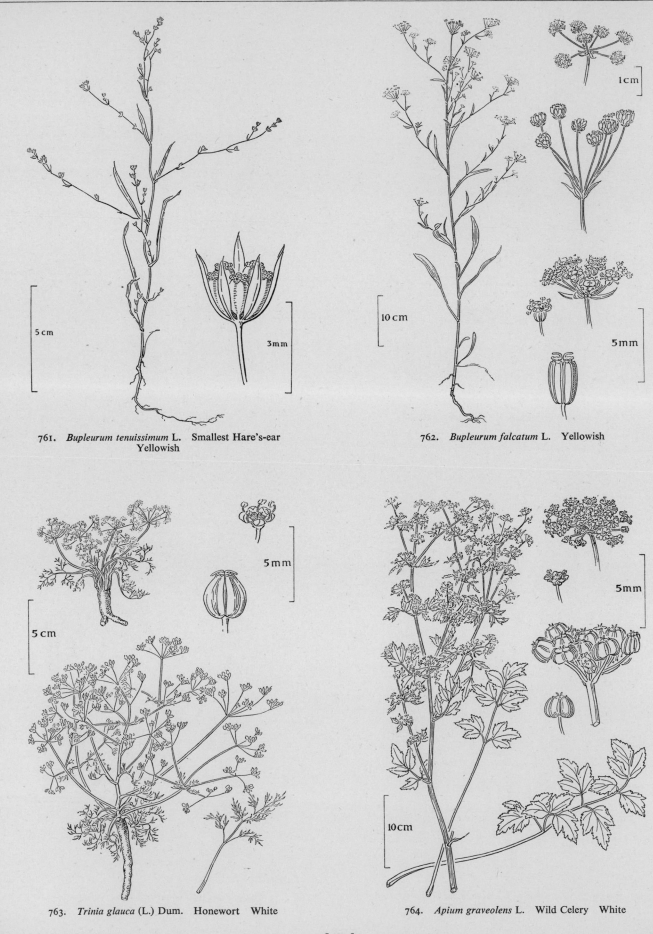

761. *Bupleurum tenuissimum* L. Smallest Hare's-ear
       Yellowish

762. *Bupleurum falcatum* L. Yellowish

763. *Trinia glauca* (L.) Dum. Honewort White

764. *Apium graveolens* L. Wild Celery White

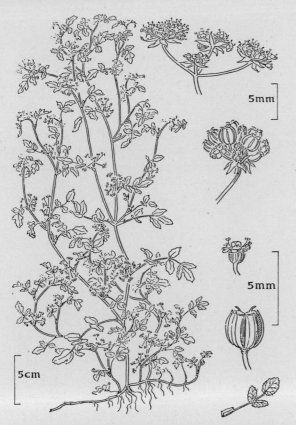

765. *Apium nodiflorum* (L.) Lag. Fool's Watercress White

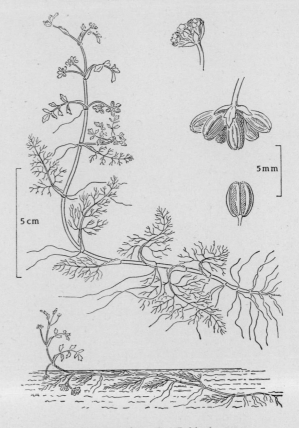

766. *Apium Inundatum* (L.) Rchb. f. White

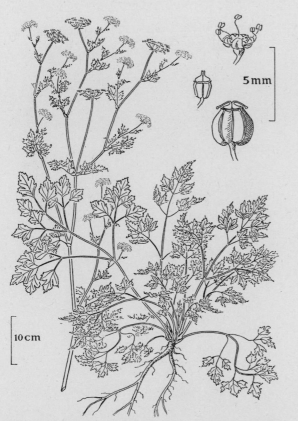

767. *Petroselinum crispum* (Mill.) Nym. Parsley Yellowish

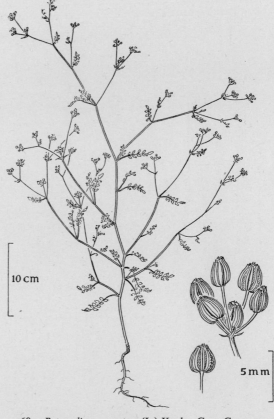

768. *Petroselinum segetum* (L.) Koch Corn Caraway
White

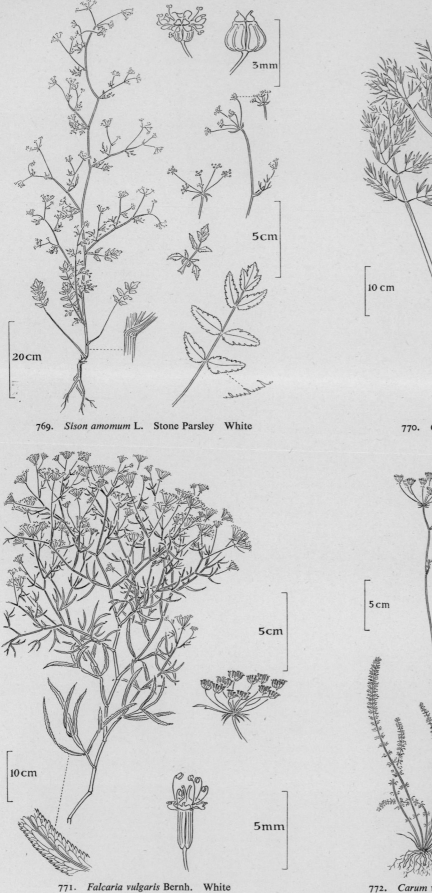

769. *Sison amomum* L. Stone Parsley White

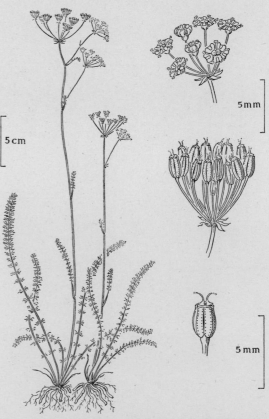

770. *Cicuta virosa* L. Cowbane White

771. *Falcaria vulgaris* Bernh. White

772. *Carum verticillatum* (L.) Koch Whorled Caraway
White

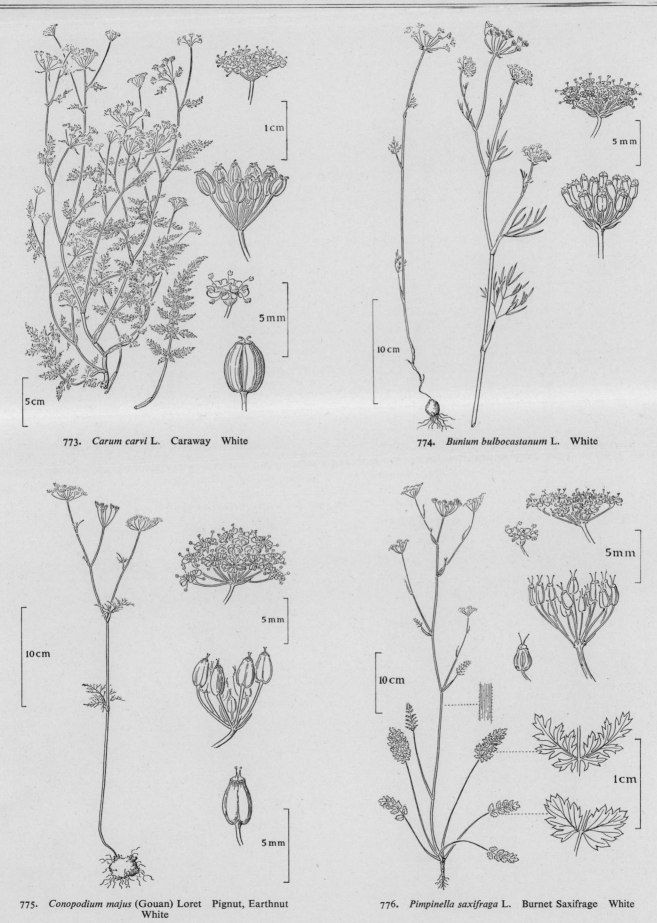

773. *Carum carvi* L. Caraway White

774. *Bunium bulbocastanum* L. White

775. *Conopodium majus* (Gouan) Loret Pignut, Earthnut White

776. *Pimpinella saxifraga* L. Burnet Saxifrage White

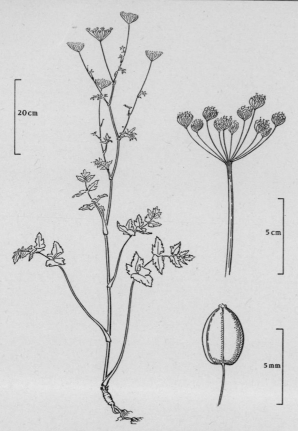

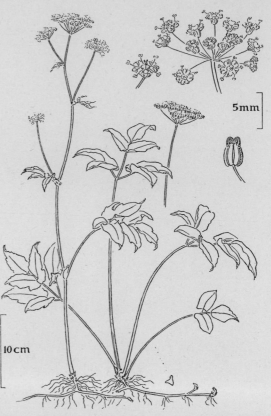

777.   *Pimpinella major* (L.) Huds.   Greater Burnet Saxifrage
White or pink

778.   *Aegopodium podagraria* L.   Goutweed, Ground Elder
White

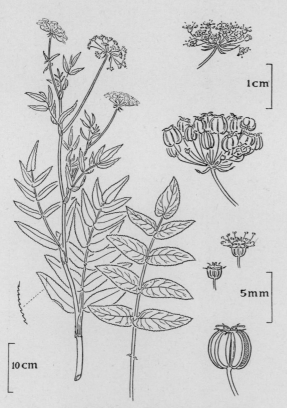

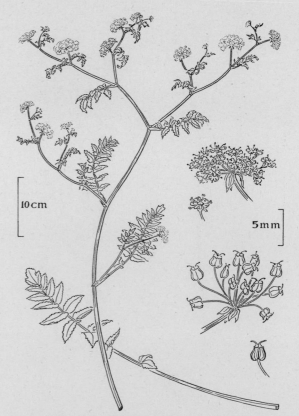

779.   *Sium latifolium* L.   Water Parsnip   White

780.   *Berula erecta* (Huds.) Coville   Narrow-leaved Water
Parsnip   White

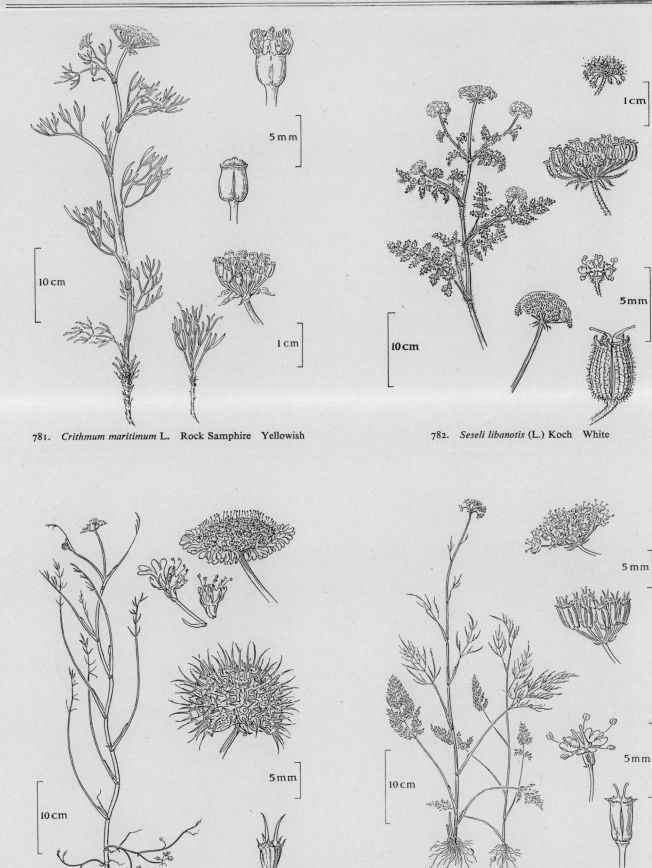

781. *Crithmum maritimum* L. Rock Samphire Yellowish

782. *Seseli libanotis* (L.) Koch White

783. *Oenanthe fistulosa* L. Water Dropwort White

784. *Oenanthe pimpinelloides* L. White

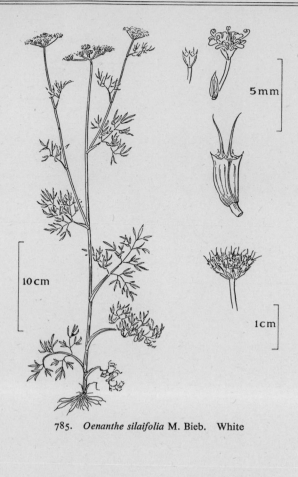

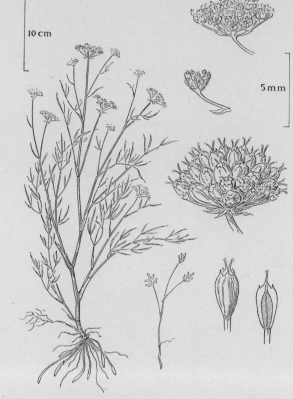

785. *Oenanthe silaifolia* M. Bieb. White

786. *Oenanthe lachenalii* C. C. Gmel. White

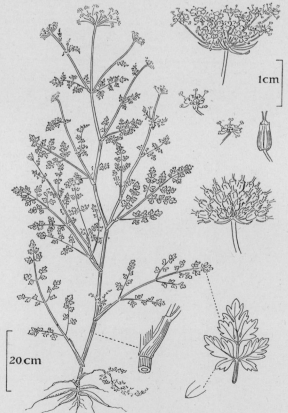

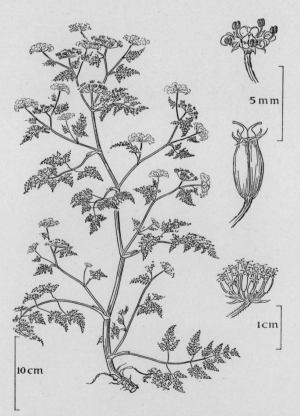

787. *Oenanthe crocata* L. Hemlock Water Dropwort
      White

788. *Oenanthe aquatica* (L.) Poir. Fine-leaved Water
      Dropwort White

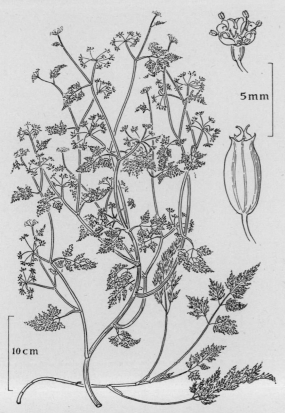

789.   *Oenanthe fluviatilis* (Bab.) Colem.   White

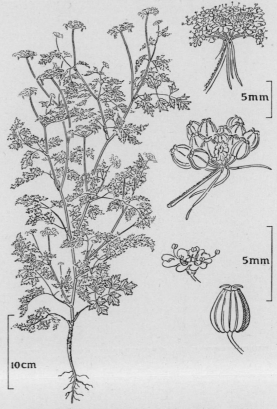

790.   *Aethusa cynapium* L.   Fool's Parsley   White

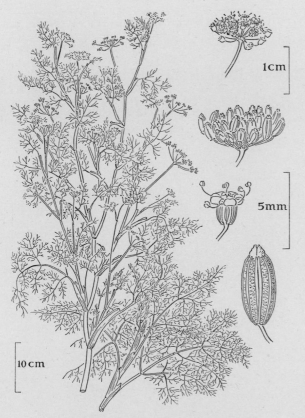

791.   *Foeniculum vulgare* Mill.   Fennel   Yellow

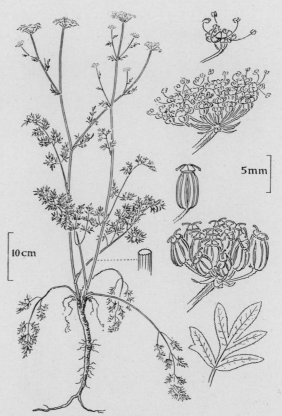

792.   *Silaum silaus* (L.) Schinz & Thell.   Pepper Saxifrage
          Yellowish

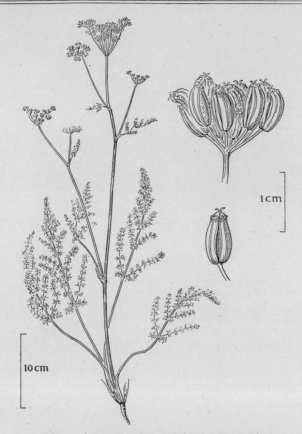

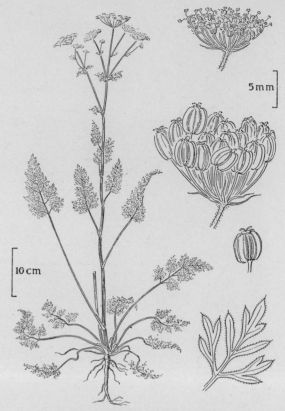

793. *Meum athamanticum* Jacq. Spignel, Meu, Baldmoney
White

794. *Selinum carvifolia* L. White

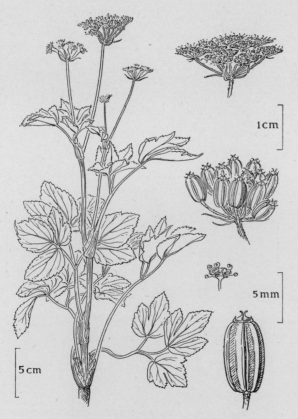

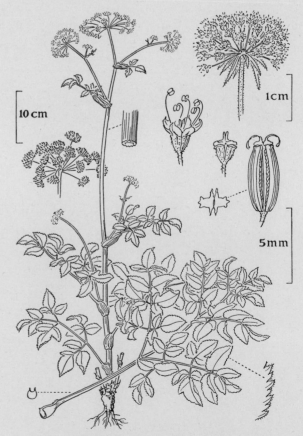

795. *Ligusticum scoticum* L. Lovage White

796. *Angelica sylvestris* L. Wild Angelica White

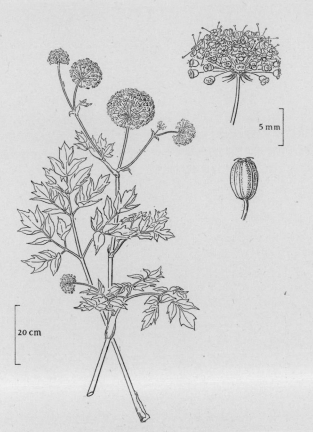

797. *Angelica archangelica* L. Angelica   Green

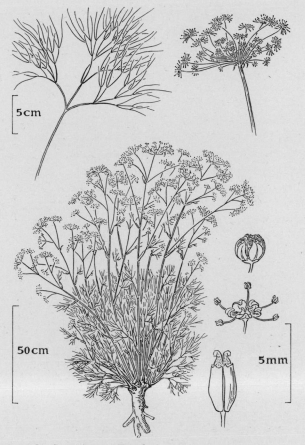

798. *Peucedanum officinale* L. Hog's Fennel, Sulphur-
weed   Yellow

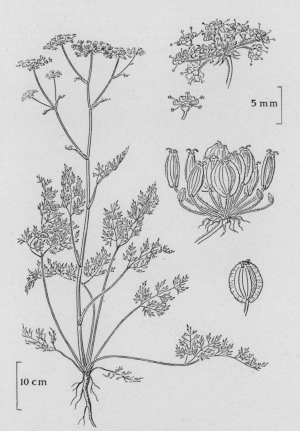

799. *Peucedanum palustre* (L.) Moench   Hog's Fennel,
Milk Parsley   White

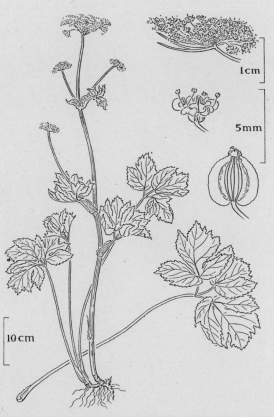

800. *Peucedanum ostruthium* (L.) Koch   Masterwort   White

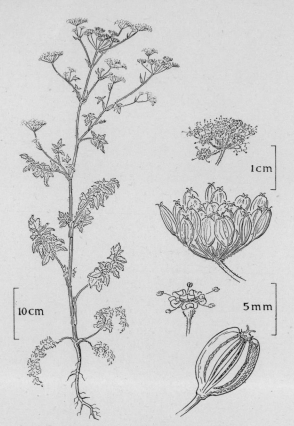

801. *Pastinaca sativa* L. Wild Parsnip Yellow

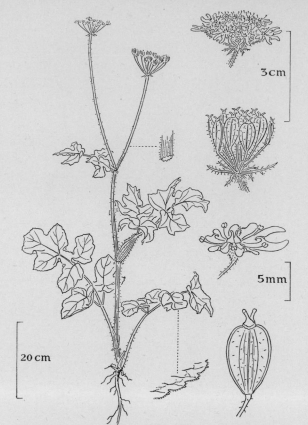

802. *Heracleum sphondylium* L. Cow Parsnip, Hogweed,
Keck Whitish

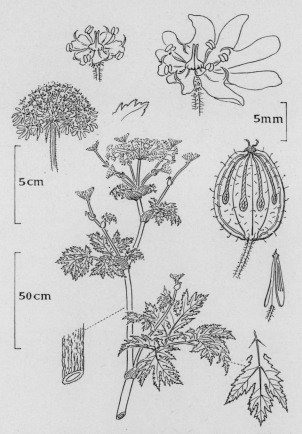

803. *Heracleum mantegazzianum* Somm. & Lev. White

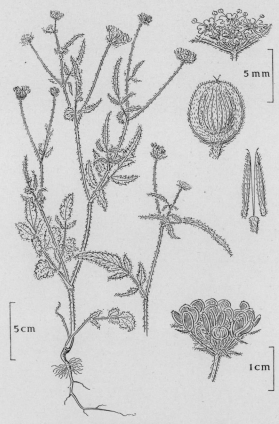

804. *Tordylium maximum* L. White

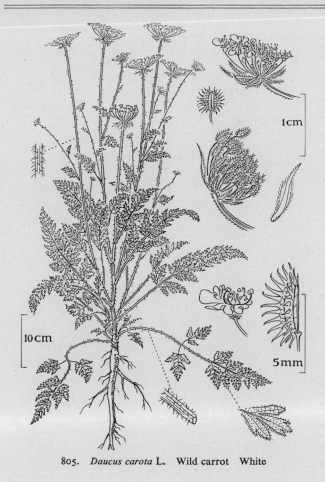

805. *Daucus carota* L. Wild carrot White

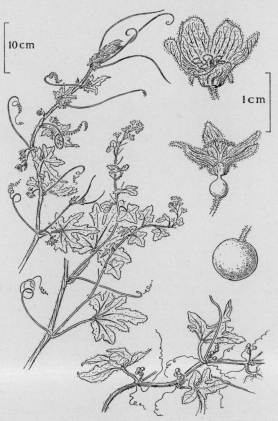

806. *Bryonia dioica* Jacq. White Bryony Greenish

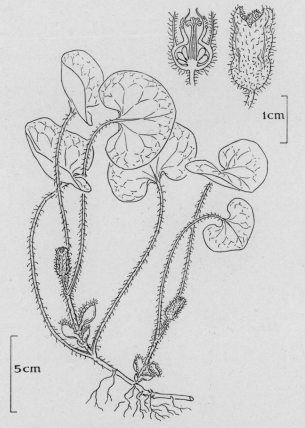

807. *Asarum europaeum* L. Asarabacca Brown

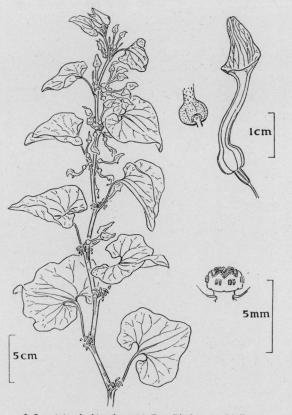

808. *Aristolochia clematitis* L. Birthwort Yellow

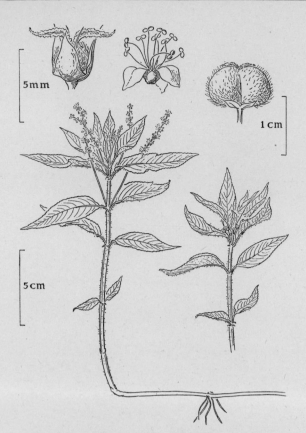

809. *Mercurialis perennis* L.   Dog's Mercury   Green

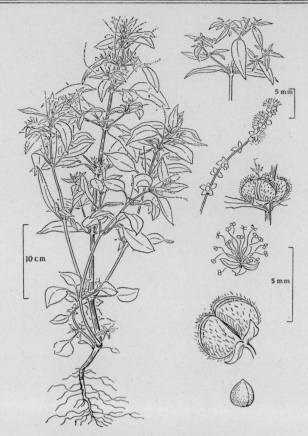

810. *Mercurialis annua* L.   Annual Mercury   Green

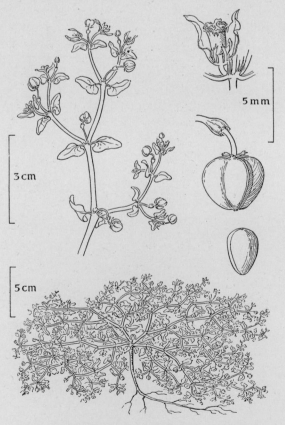

811. *Euphorbia peplis* L.   Purple Spurge   Greenish

812. *Euphorbia lathyrus* L.   Caper Spurge   Greenish

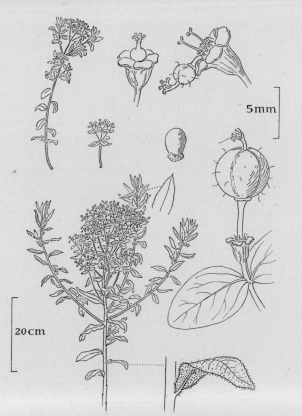

5mm

20cm

813. *Euphorbia pilosa* L.   Hairy Spurge   Greenish

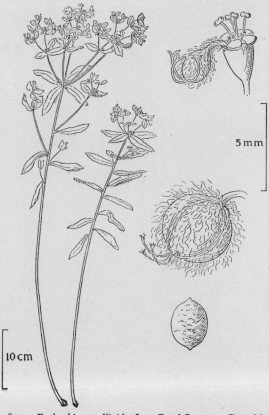

5mm

10cm

814. *Euphorbia corallioides* L.   Coral Spurge   Greenish

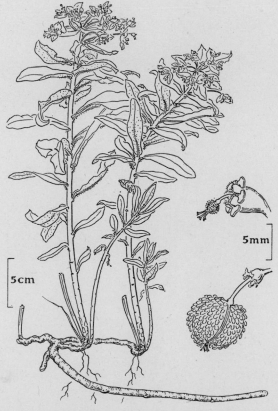

5mm

5cm

815. *Euphorbia hyberna* L.   Irish Spurge   Greenish

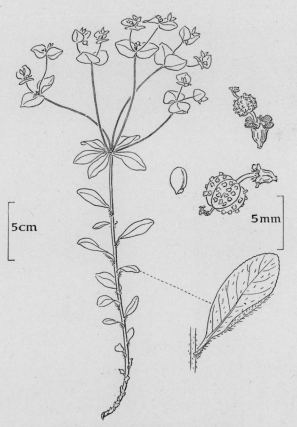

5cm

5mm

816. *Euphorbia dulcis* L.   Greenish

5-2

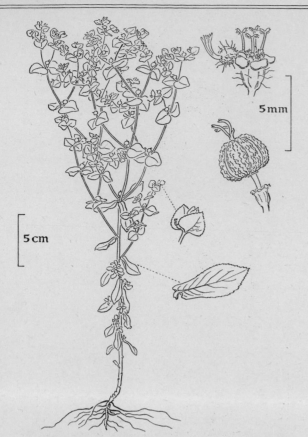

817. *Euphorbia platyphyllos* L. Broad Spurge   Greenish

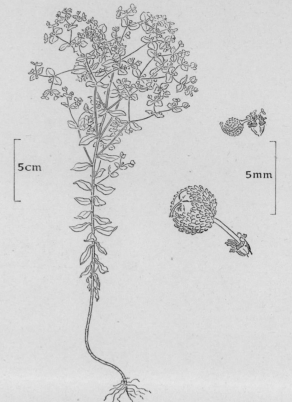

818. *Euphorbia stricta* L. Upright Spurge   Greenish

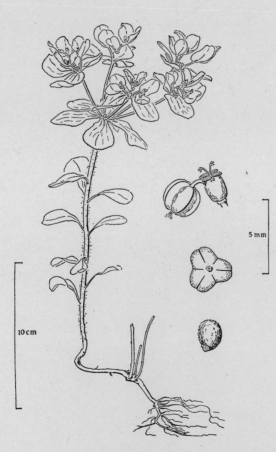

819. *Euphorbia helioscopia* L. Sun Spurge   Greenish

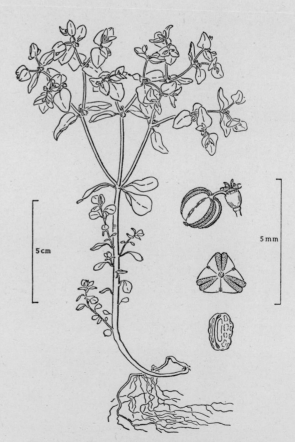

820. *Euphorbia peplus* L. Petty Spurge   Greenish

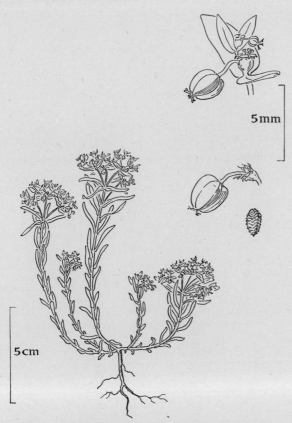

821.   *Euphorbia exigua* L.   Dwarf Spurge   Greenish

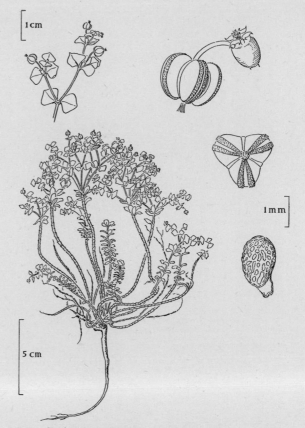

822.   *Euphorbia portlandica* L.   Portland Spurge   Greenish

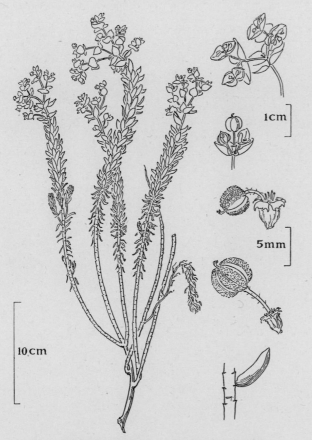

823.   *Euphorbia paralias* L.   Sea Spurge   Greenish

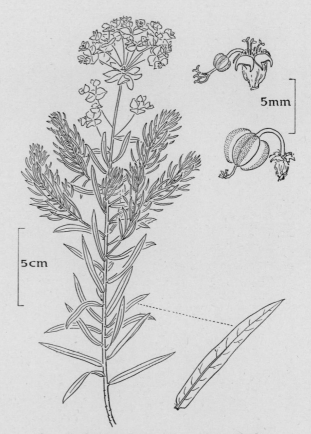

824.   *Euphorbia uralensis* Fisch. ex Link   Greenish

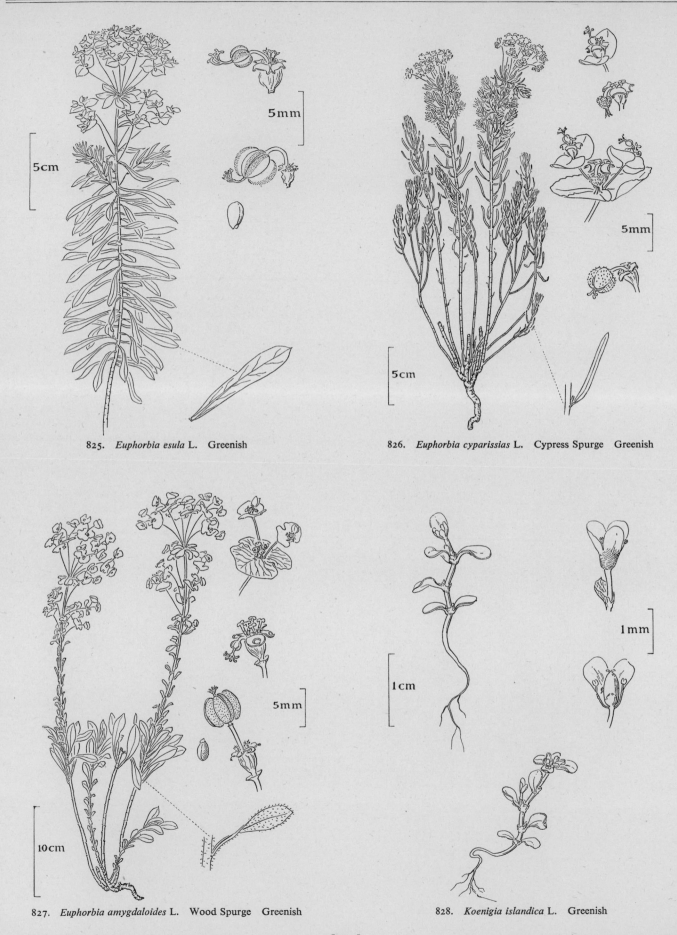

825. *Euphorbia esula* L.   Greenish

826. *Euphorbia cyparissias* L.   Cypress Spurge   Greenish

827. *Euphorbia amygdaloides* L.   Wood Spurge   Greenish

828. *Koenigia islandica* L.   Greenish

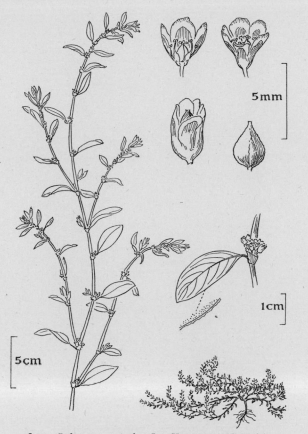

829. *Polygonum aviculare* L. Knotgrass Greenish

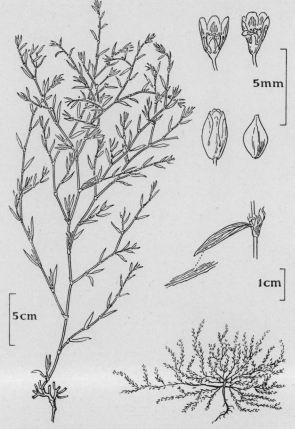

830. *Polygonum rurivagum* Jord. ex Bor. Greenish

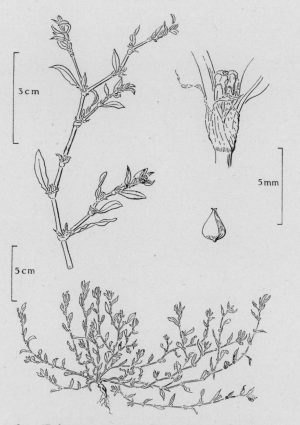

831. *Polygonum arenastrum* Bor. Knotgrass Greenish

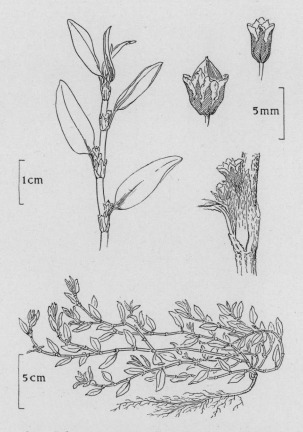

832. *Polygonum raii* Bab. Ray's Knotgrass Greenish

[71]

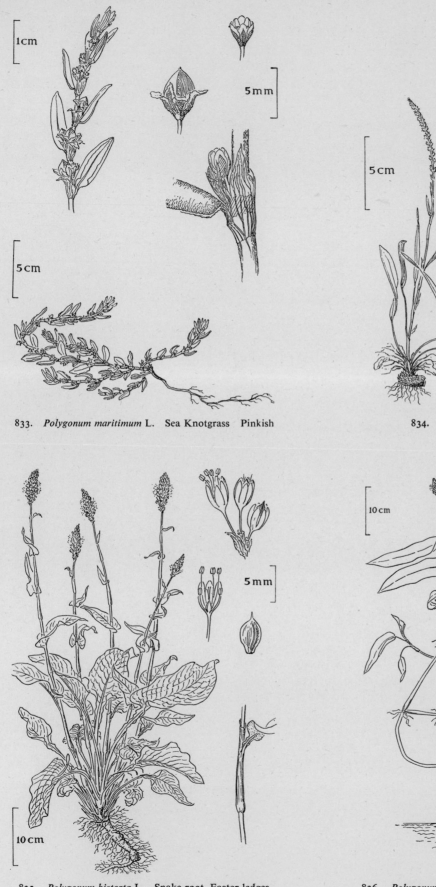

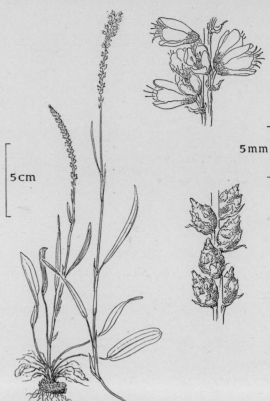

833. *Polygonum maritimum* L.   Sea Knotgrass   Pinkish

834. *Polygonum viviparum* L.   White

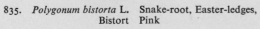

835. *Polygonum bistorta* L.   Snake-root, Easter-ledges,
Bistort   Pink

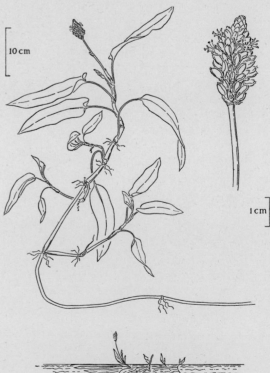

836. *Polygonum amphibium* L.   Amphibious Bistort   Pink
(Water form)

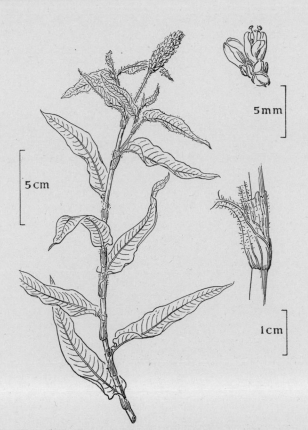

837. *Polygonum amphibium* L.   Amphibious Bistort   Pink
(Land form)

838. *Polygonum persicaria* L.   Persicaria   Pink

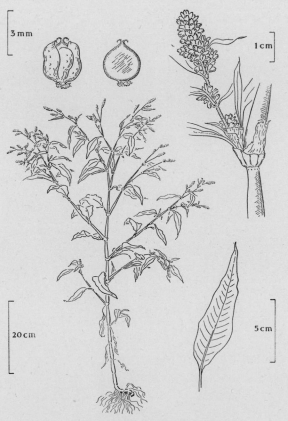

839. *Polygonum lapathifolium* L.   Pale Persicaria
Greenish-white

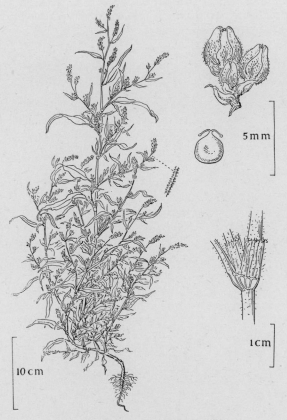

840. *Polygonum nodosum* Pers.   Pink

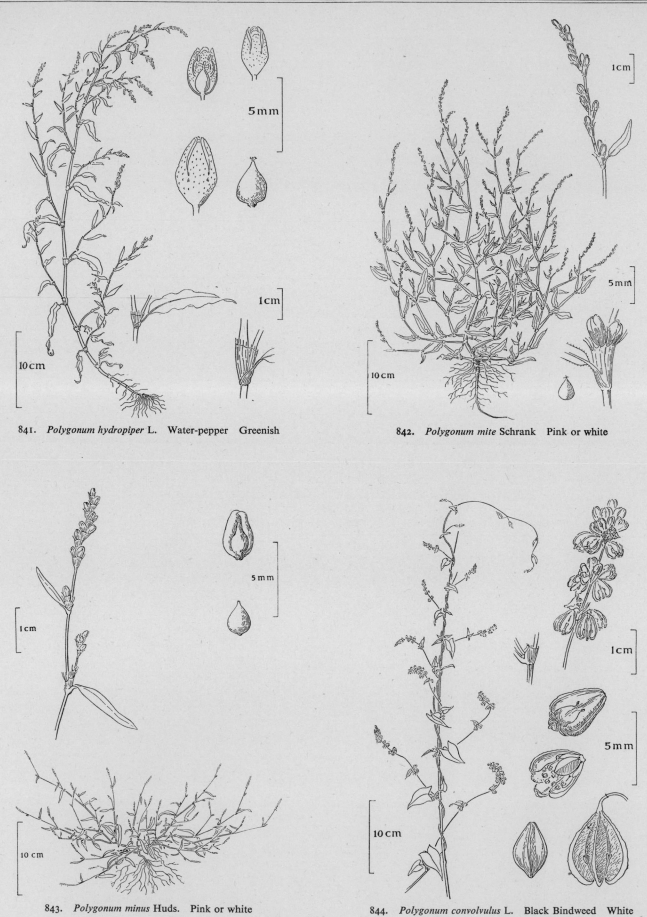

841.   *Polygonum hydropiper* L.   Water-pepper   Greenish

842.   *Polygonum mite* Schrank   Pink or white

843.   *Polygonum minus* Huds.   Pink or white

844.   *Polygonum convolvulus* L.   Black Bindweed   White

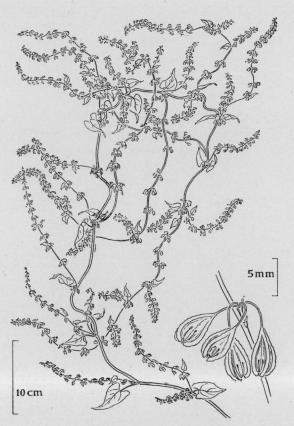

5mm

10 cm

845. *Polygonum dumetorum* L.   White

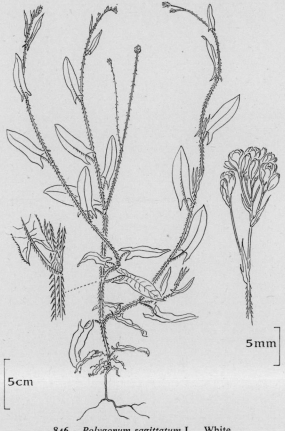

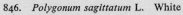

5mm

5cm

846. *Polygonum sagittatum* L.   White

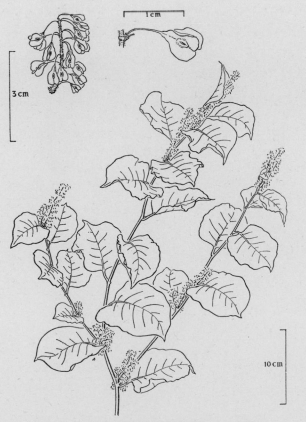

1 cm

3 cm

10 cm

847. *Polygonum cuspidatum* Sieb. & Zucc.   White

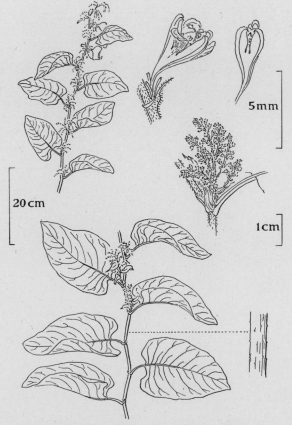

5mm

20 cm

1 cm

848. *Polygonum sachalinense* F. Schmidt   White

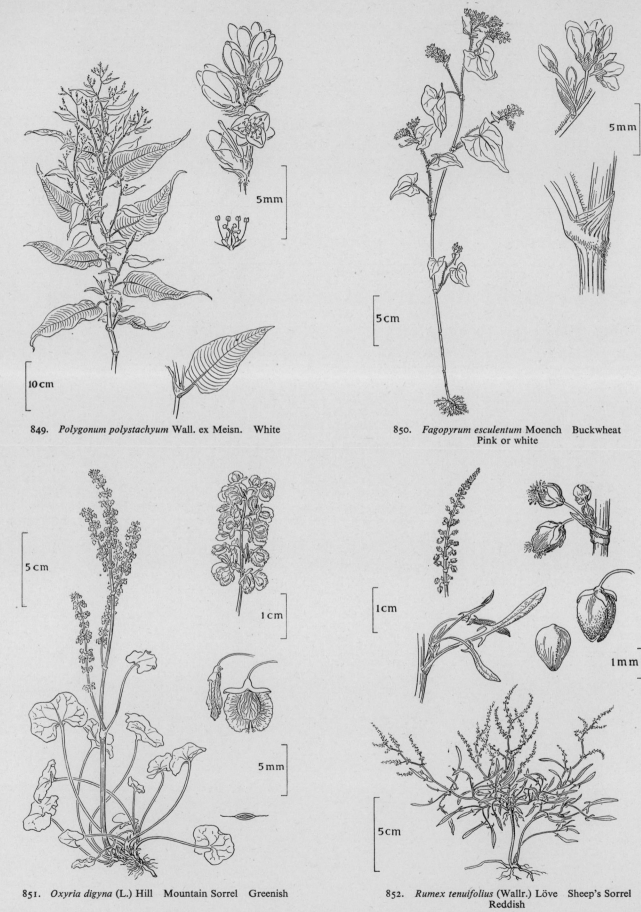

849.   *Polygonum polystachyum* Wall. ex Meisn.   White

850.   *Fagopyrum esculentum* Moench   Buckwheat
Pink or white

851.   *Oxyria digyna* (L.) Hill   Mountain Sorrel   Greenish

852.   *Rumex tenuifolius* (Wallr.) Löve   Sheep's Sorrel
Reddish

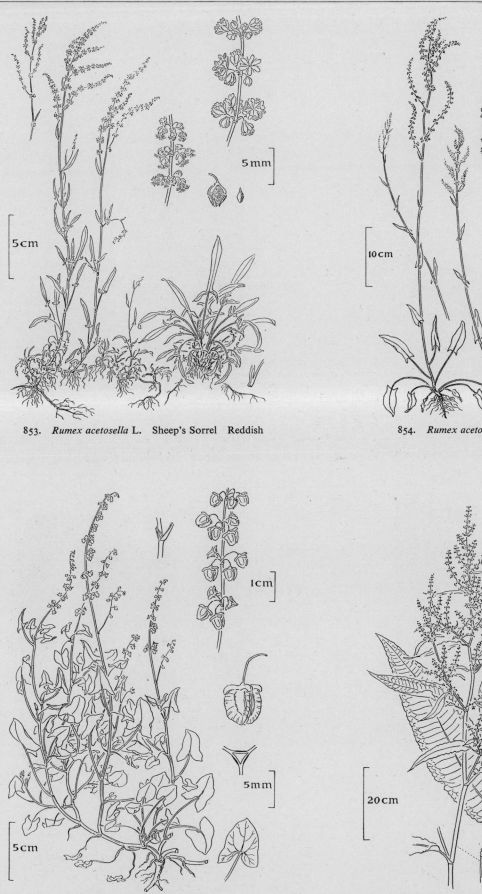

853. *Rumex acetosella* L.   Sheep's Sorrel   Reddish

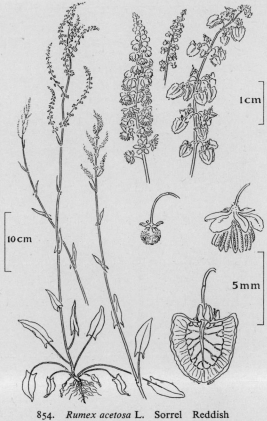

854. *Rumex acetosa* L.   Sorrel   Reddish

855. *Rumex scutatus* L.   Reddish

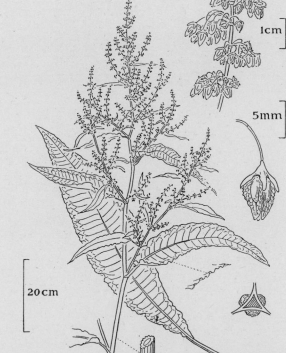

856. *Rumex hydrolapathum* Huds.   Great Water Dock
Greenish

857. *Rumex alpinus* L.   Monk's Rhubarb   Greenish

858. *Rumex aquaticus* L.   Greenish

859. *Rumex longifolius* DC.   Greenish

860. *Rumex cristatus* DC.   Greenish

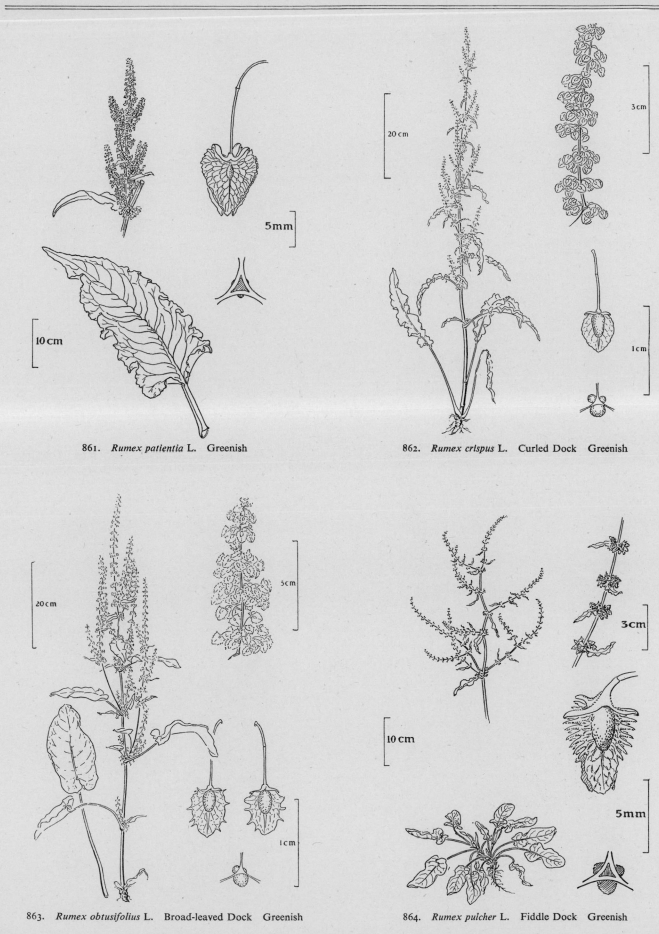

861. *Rumex patientia* L.   Greenish

862. *Rumex crispus* L.   Curled Dock   Greenish

863. *Rumex obtusifolius* L.   Broad-leaved Dock   Greenish

864. *Rumex pulcher* L.   Fiddle Dock   Greenish

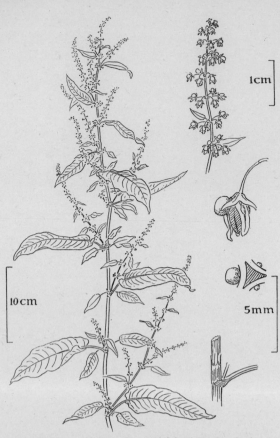

1cm

10cm

5mm

865. *Rumex sanguineus* L. Red-veined Dock Greenish

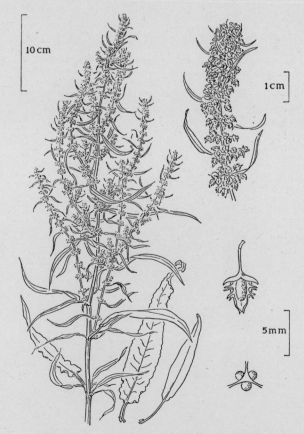

1cm

5mm

10cm

866. *Rumex conglomeratus* Murr. Sharp Dock Greenish

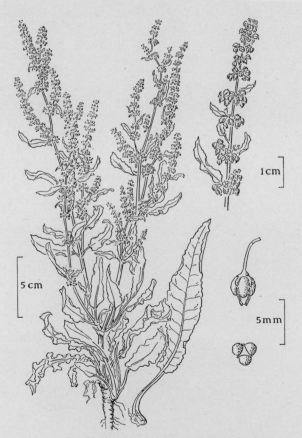

1cm

5cm

5mm

867. *Rumex rupestris* Le Gall Shore Dock Greenish

10cm

1cm

5mm

868. *Rumex palustris* Sm. Marsh Dock Greenish

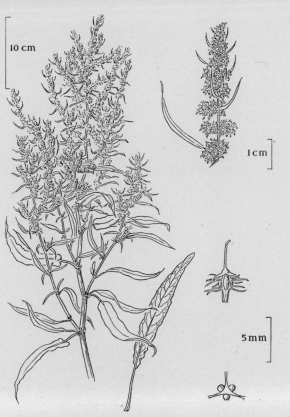

869. *Rumex maritimus* L.   Golden Dock   Greenish

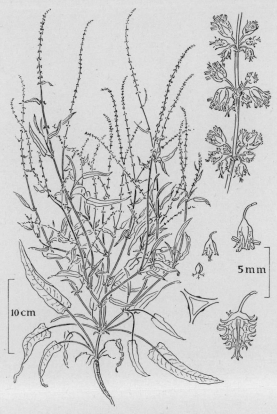

870. *Rumex brownii* Campd.   Greenish

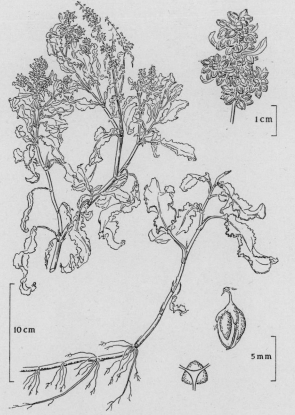

871. *Rumex frutescens* Thou.   Greenish

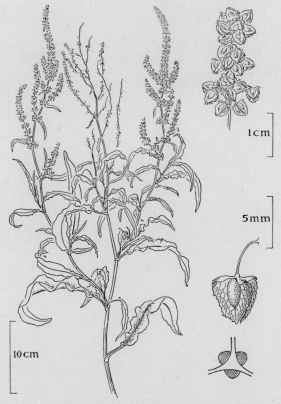

872. *Rumex triangulivalvis* (Danser) Rech. f.   Green

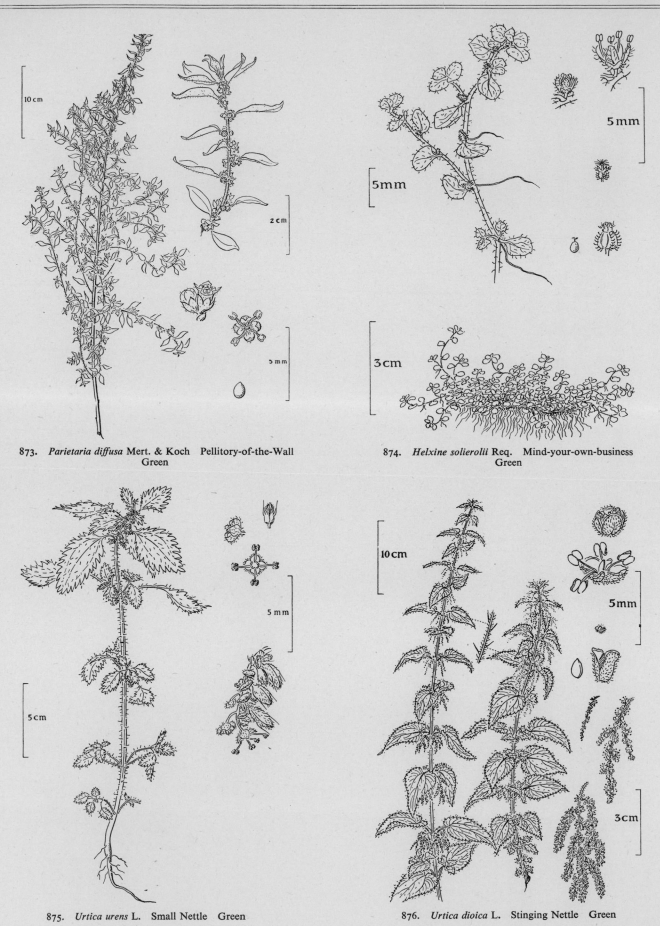

873.   *Parietaria diffusa* Mert. & Koch   Pellitory-of-the-Wall
        Green

874.   *Helxine solierolii* Req.   Mind-your-own-business
        Green

875.   *Urtica urens* L.   Small Nettle   Green

876.   *Urtica dioica* L.   Stinging Nettle   Green

877. *Humulus lupulus* L.   Hop   Green

878. *Ulmus glabra* Huds.   Wych Elm   Reddish

879. *Ulmus procera* Salisb.   English Elm   Reddish

880. *Ulmus carpinifolia* Gleditsch   Smooth Elm   Reddish

6-2

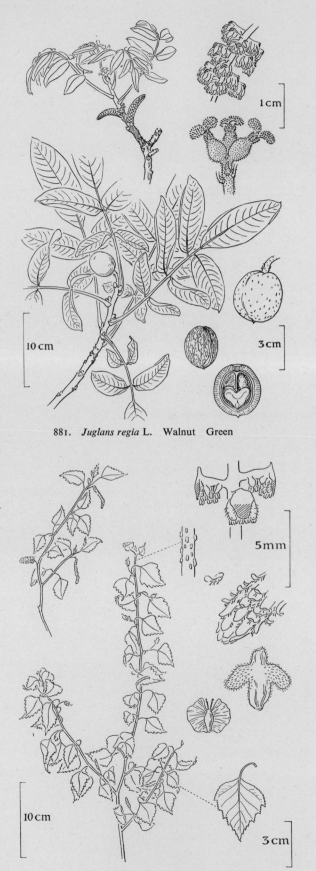

881.   *Juglans regia* L.   Walnut   Green

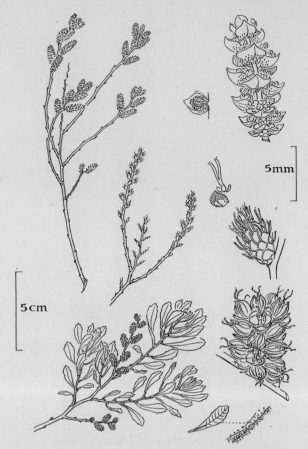

882.   *Myrica gale* L.   Bog Myrtle, Sweet Gale   Brown

883.   *Betula pendula* Roth.   Silver Birch   Greenish

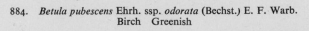

884.   *Betula pubescens* Ehrh. ssp. *odorata* (Bechst.) E. F. Warb.
Birch   Greenish

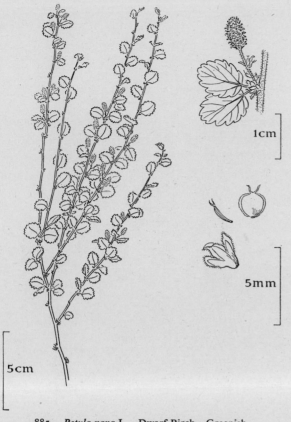

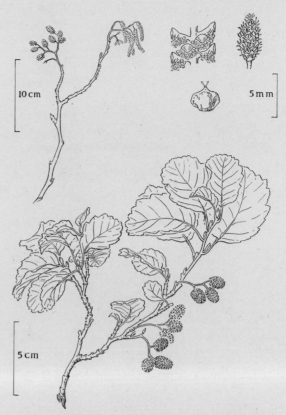

885. *Betula nana* L. Dwarf Birch Greenish

886. *Alnus glutinosa* (L.) Gaertn. Alder Greenish

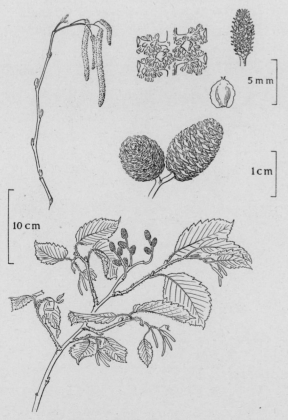

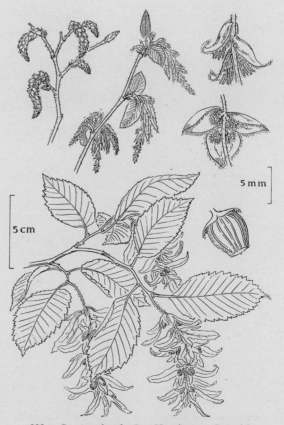

887. *Alnus incana* (L.) Moench Grey Alder Greenish

888. *Carpinus betulus* L. Hornbeam Greenish

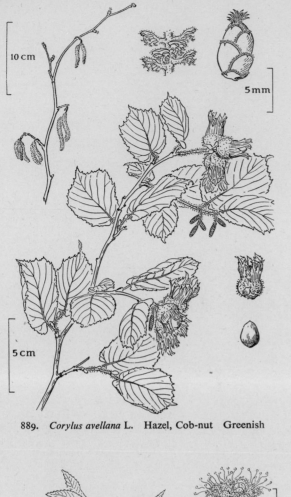

889. *Corylus avellana* L. Hazel, Cob-nut Greenish

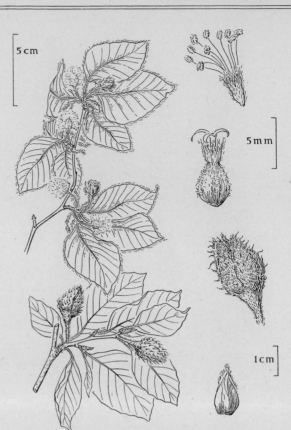

890. *Fagus sylvatica* L. Beech Greenish

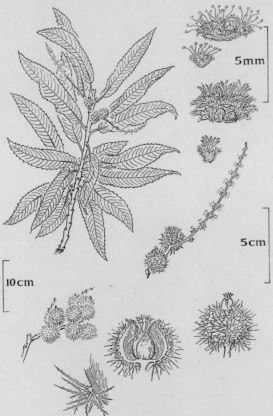

891. *Castanea sativa* L. Sweet or Spanish Chestnut
            Yellowish

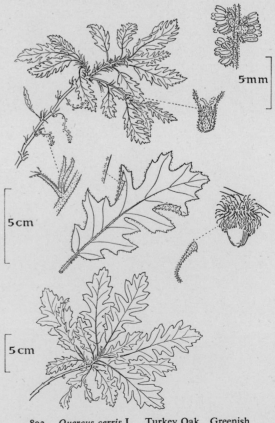

892. *Quercus cerris* L. Turkey Oak Greenish

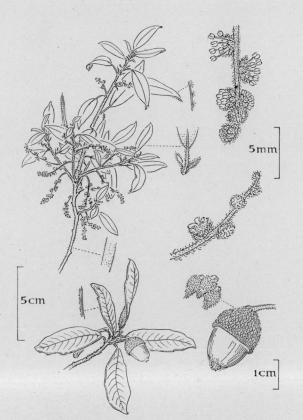

893. *Quercus ilex* L.   Evergreen or Holm Oak   Greenish

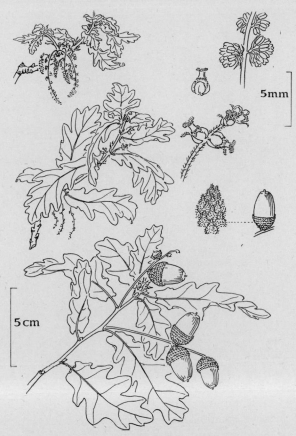

894. *Quercus robur* L.   Common or Pedunculate Oak
      Greenish

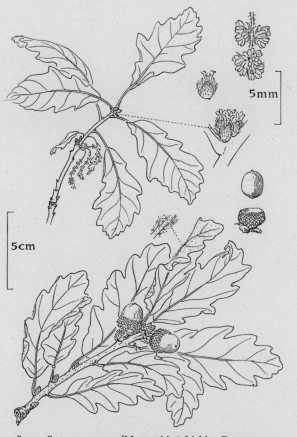

895. *Quercus petraea* (Mattuschka) Liebl.   Durmast or
      Sessile Oak   Greenish

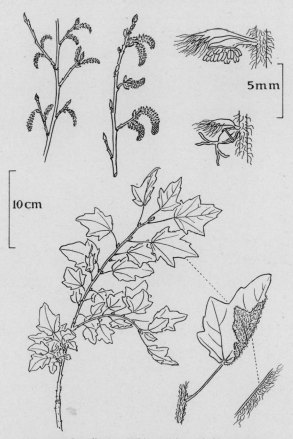

896. *Populus alba* L.   White Poplar, Abele   Greenish
      or purplish

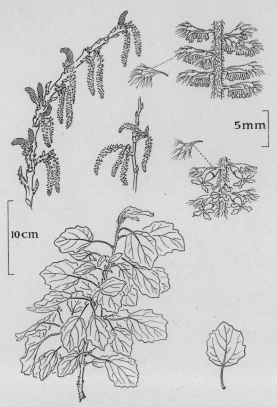

897. *Populus canescens* Sm.   Grey Poplar   Greenish
or purplish

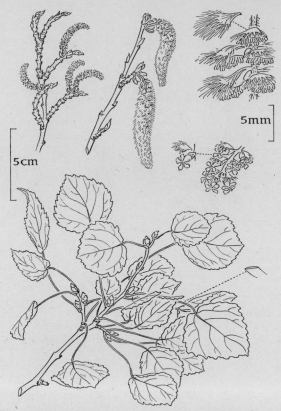

898. *Populus tremula* L.   Aspen   Purplish

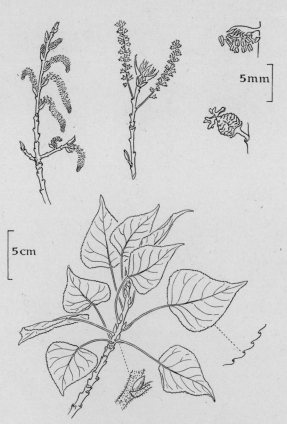

899. *Populus nigra* L.   Black Poplar   Reddish

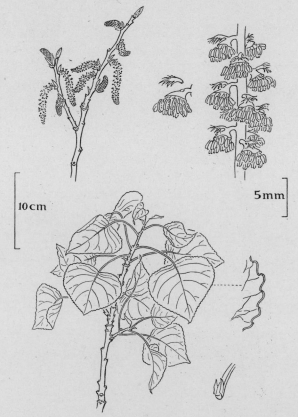

900. *Populus × canadensis* Moench   Black Italian Poplar
Reddish

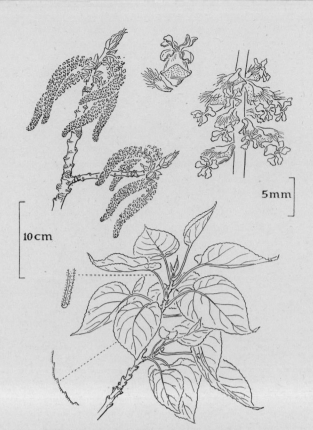

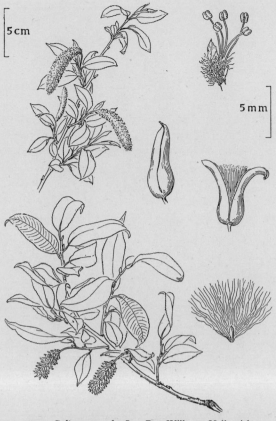

901. *Populus gileadensis* Rouleau Balm of Gilead Greenish

902. *Salix pentandra* L. Bay Willow Yellowish

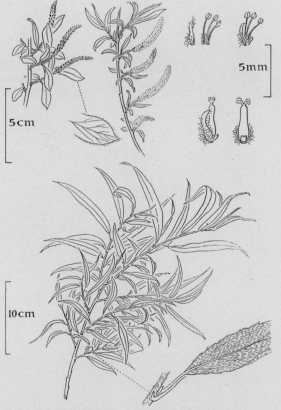

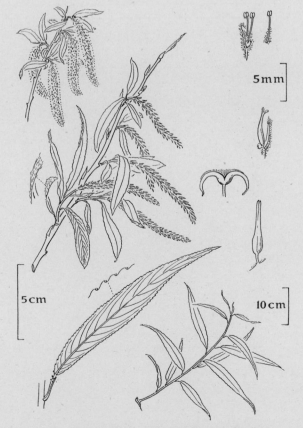

903. *Salix alba* L. White Willow Yellowish

904. *Salix fragilis* L. Crack Willow Yellowish

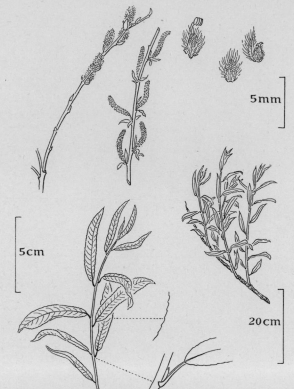

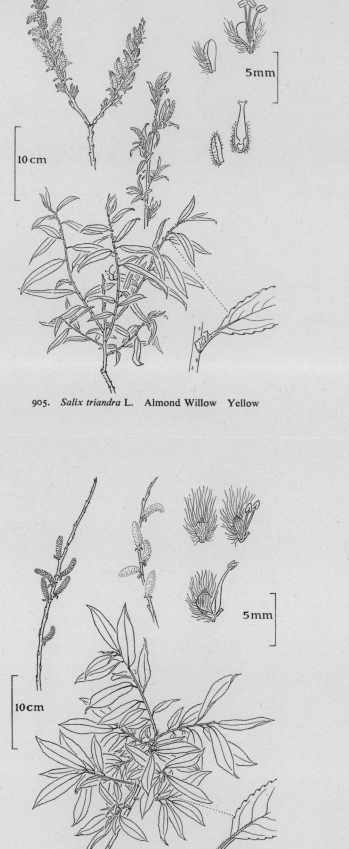

905. *Salix triandra* L. Almond Willow Yellow

906. *Salix purpurea* L. Purple Willow Purplish

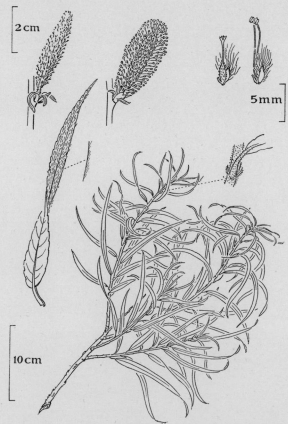

907. *Salix daphnoides* Vill. Yellowish or greyish

908. *Salix viminalis* L. Common Osier Yellowish or greyish

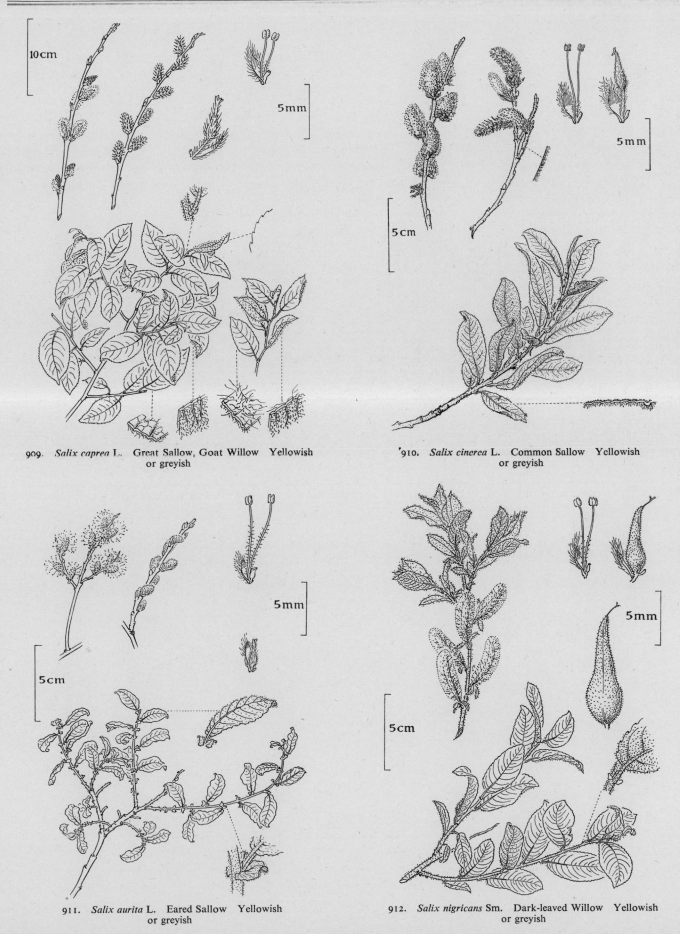

909. *Salix caprea* L.   Great Sallow, Goat Willow   Yellowish
       or greyish

'910. *Salix cinerea* L.   Common Sallow   Yellowish
       or greyish

911. *Salix aurita* L.   Eared Sallow   Yellowish
       or greyish

912. *Salix nigricans* Sm.   Dark-leaved Willow   Yellowish
       or greyish

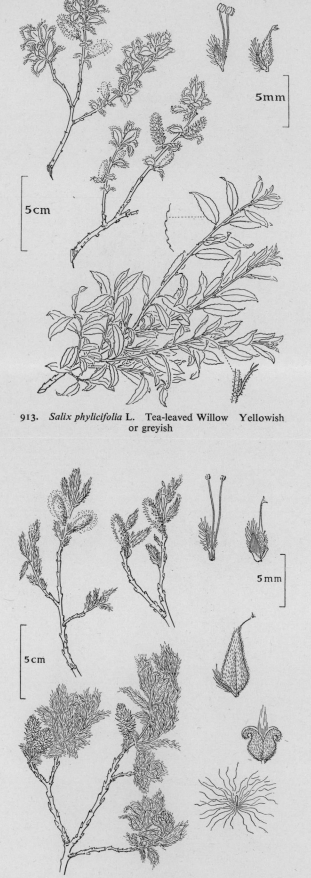

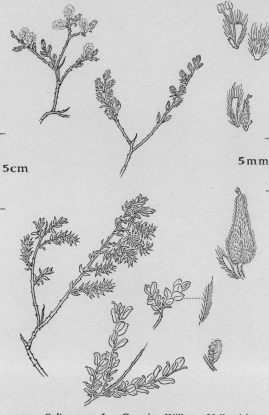

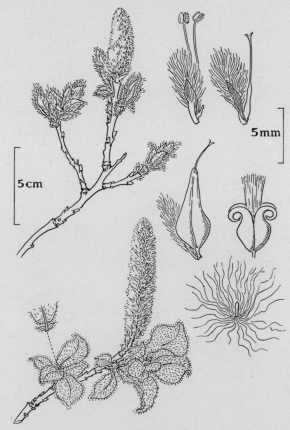

913.　*Salix phylicifolia* L.　Tea-leaved Willow　Yellowish
　　　or greyish

914.　*Salix repens* L.　Creeping Willow　Yellowish
　　　or greyish

915.　*Salix lapponum* L.　Downy Willow　Yellowish
　　　or greyish

916.　*Salix lanata* L.　Woolly Willow　Yellowish

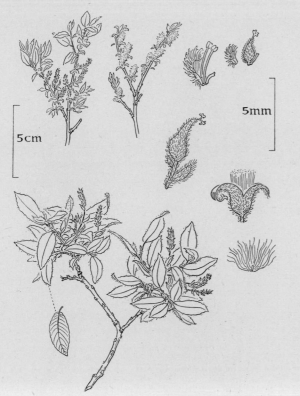

917.   *Salix arbuscula* L.   Yellowish or greyish

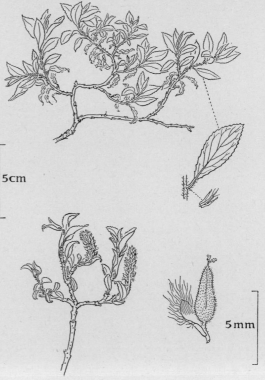

918.   *Salix myrsinites* L.   Purplish

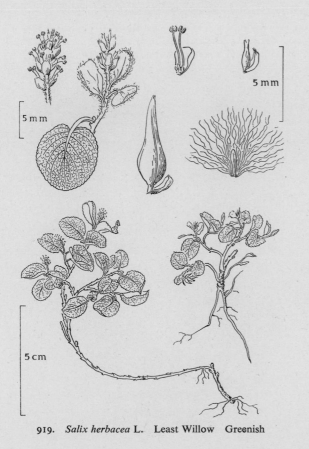

919.   *Salix herbacea* L.   Least Willow   Greenish

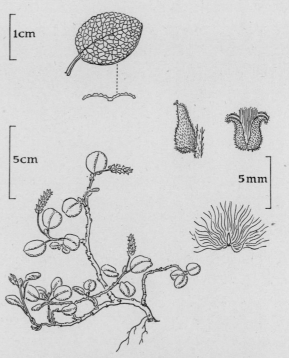

920.   *Salix reticulata* L.   Reticulate Willow   Greyish

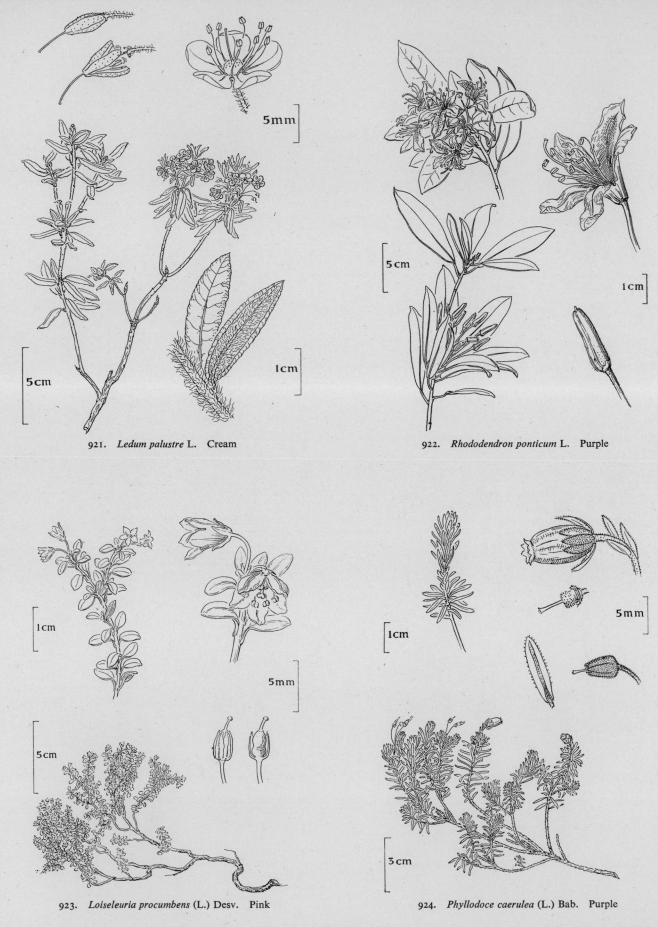

921. *Ledum palustre* L. Cream

922. *Rhododendron ponticum* L. Purple

923. *Loiseleuria procumbens* (L.) Desv. Pink

924. *Phyllodoce caerulea* (L.) Bab. Purple

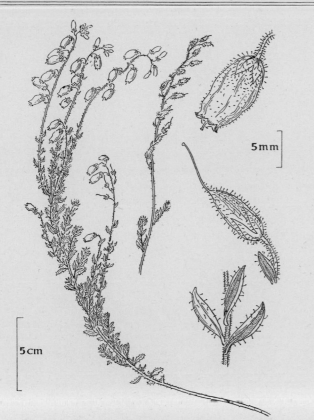

925. *Daboecia cantabrica* (Huds.) C. Koch   St Dabeoc's
         Heath   Purple

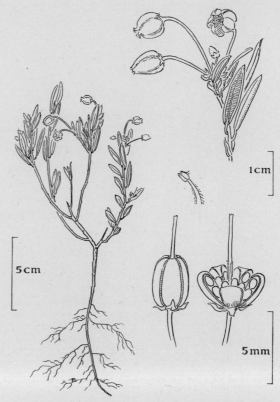

926. *Andromeda polifolia* L.   Marsh Andromeda   Pink

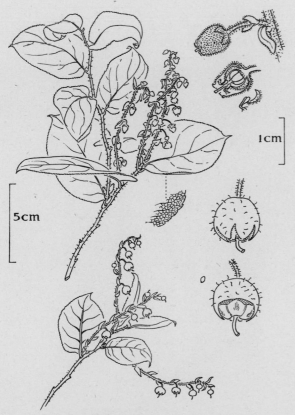

927. *Gaultheria shallon* Pursh   Pinkish-white

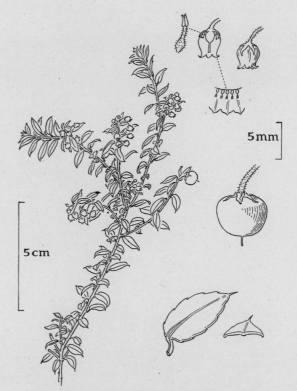

928. *Pernettya mucronata* (L. f.) Gaudich. ex Spreng.
         White

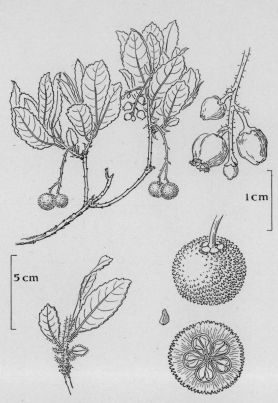

1cm

5 cm

929. *Arbutus unedo* L.   Strawberry Tree   White

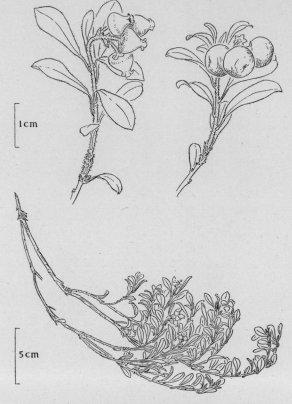

1cm

5cm

930. *Arctostaphyllos uva-ursi* (L.) Spreng.   Bearberry
Pinkish-white

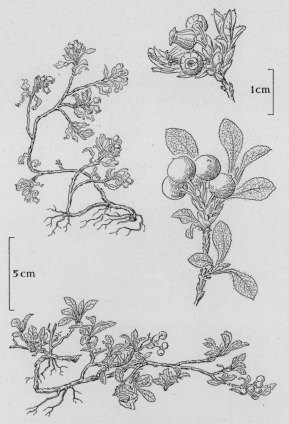

1cm

5 cm

931. *Arctous alpinus* (L.) Nied.   Black Bearberry   White

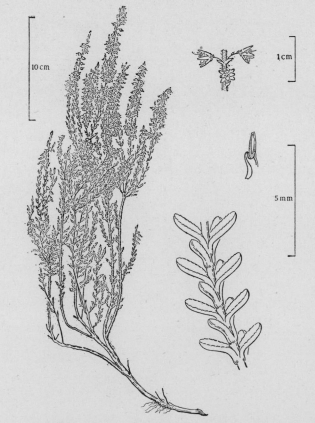

10 cm

1cm

5 mm

932. *Calluna vulgaris* (L.) Hull   Ling, Heather   Pale purple

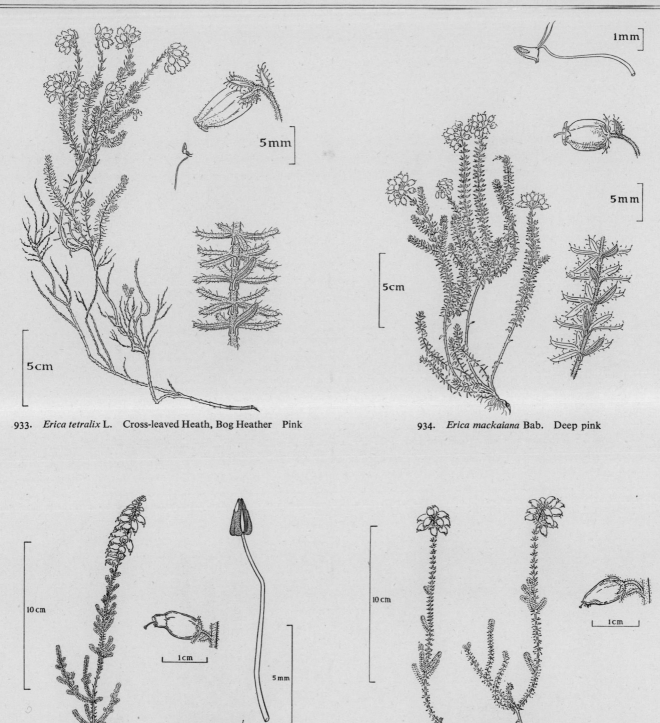

933. *Erica tetralix* L.   Cross-leaved Heath, Bog Heather   Pink

934. *Erica mackaiana* Bab.   Deep pink

935. *Erica ciliaris* L.   Dorset Heath   Deep pink

936. *Erica ciliaris × tetralix*   Pink

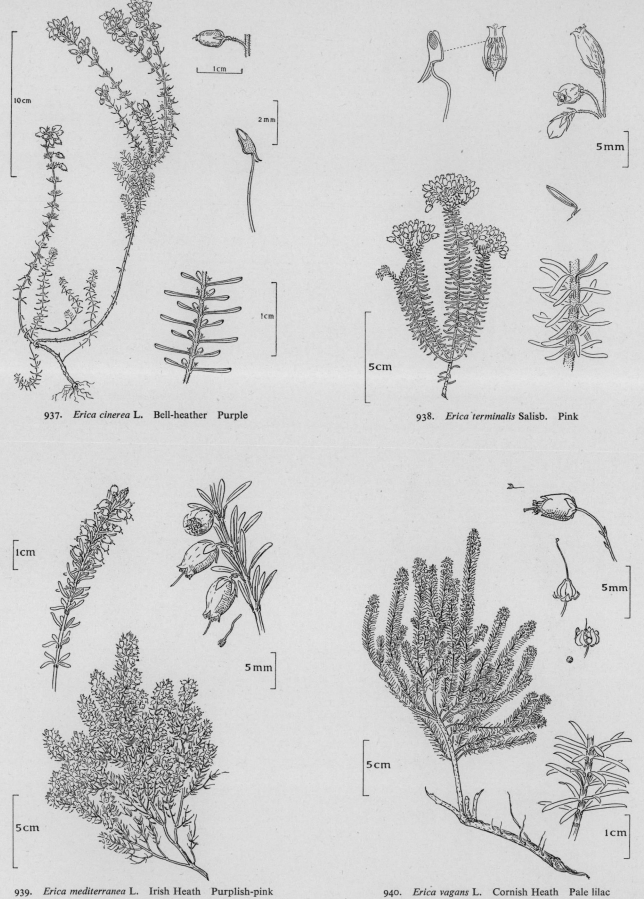

937. *Erica cinerea* L. Bell-heather Purple

938. *Erica terminalis* Salisb. Pink

939. *Erica mediterranea* L. Irish Heath Purplish-pink

940. *Erica vagans* L. Cornish Heath Pale lilac

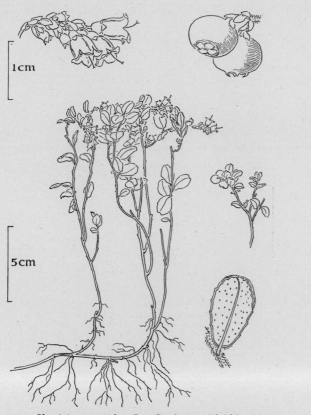

941. *Vaccinium vitis-idaea* L.   Cowberry   Pinkish-white

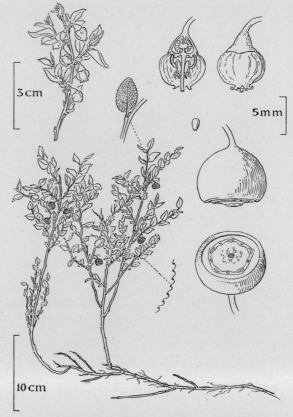

942. *Vaccinium myrtillus* L.   Bilberry, Whortleberry
Greenish-pink

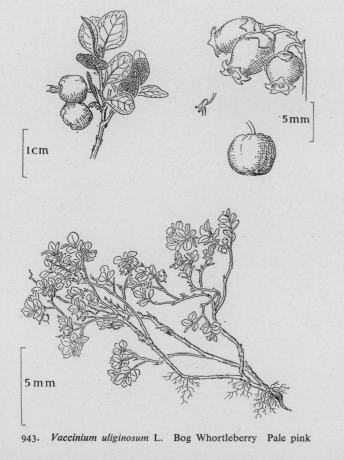

943. *Vaccinium uliginosum* L. Bog Whortleberry Pale pink

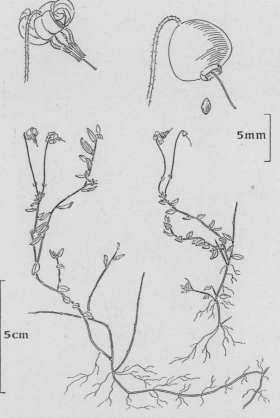

944. *Vaccinium oxycoccus* L. Cranberry Pink

7-2

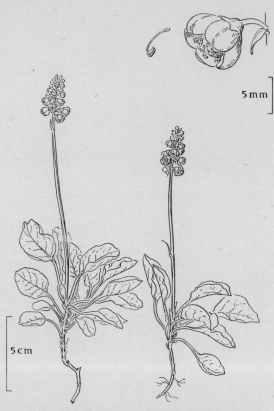

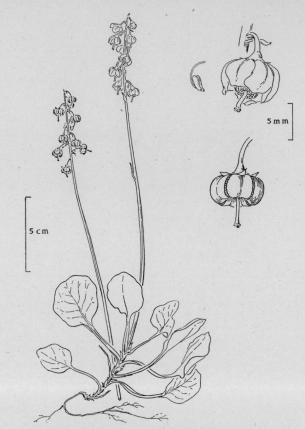

945. *Pyrola minor* L.   Common Wintergreen   Pinkish

946. *Pyrola media* Sw.   Intermediate Wintergreen   Whitish

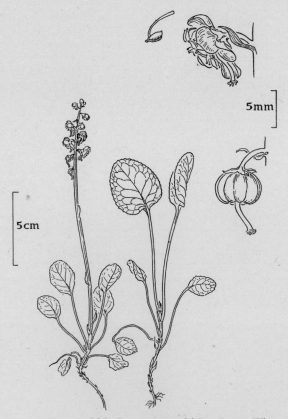

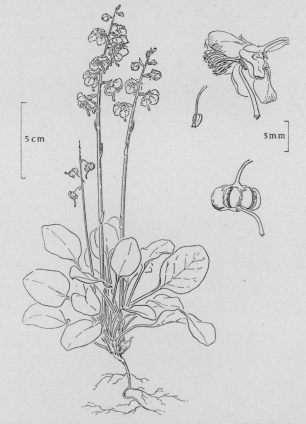

947. *Pyrola rotundifolia* L.   ssp. *rotundifolia*   Larger Winter-
      green   White

948. *Pyrola rotundifolia* ssp. *maritima* (Kenyon) E. F. Warb.
      White

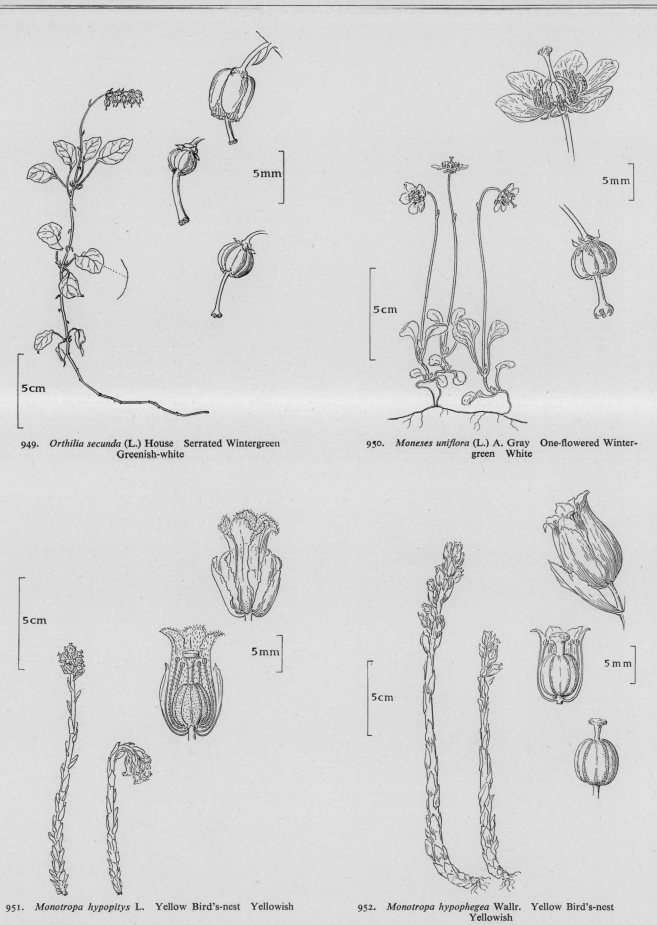

949. *Orthilia secunda* (L.) House   Serrated Wintergreen
Greenish-white

950. *Moneses uniflora* (L.) A. Gray   One-flowered Wintergreen   White

951. *Monotropa hypopitys* L.   Yellow Bird's-nest   Yellowish

952. *Monotropa hypophegea* Wallr.   Yellow Bird's-nest
Yellowish

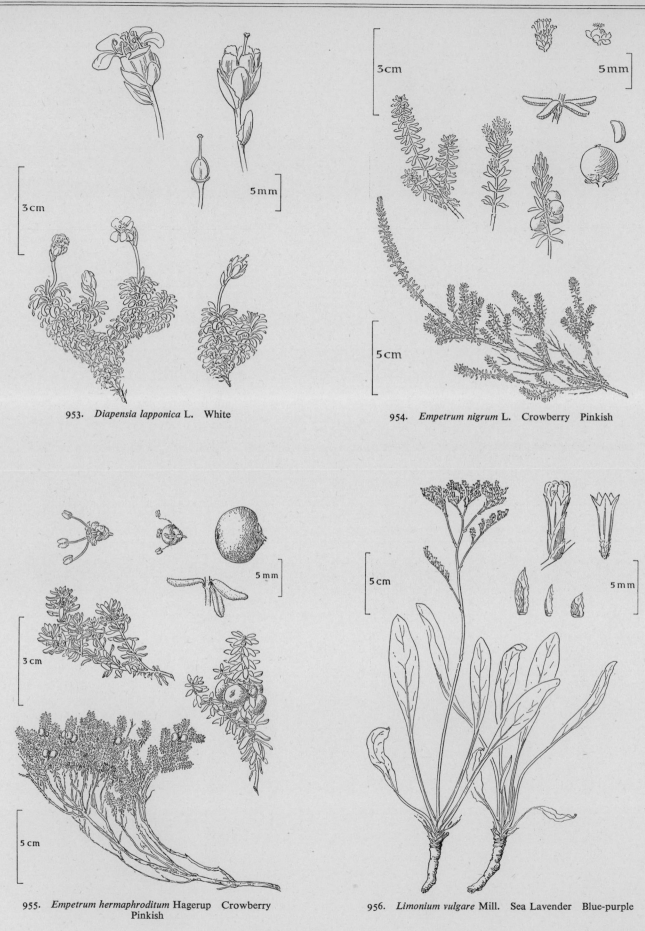

953. *Diapensia lapponica* L.   White

954. *Empetrum nigrum* L.   Crowberry   Pinkish

955. *Empetrum hermaphroditum* Hagerup   Crowberry
     Pinkish

956. *Limonium vulgare* Mill.   Sea Lavender   Blue-purple

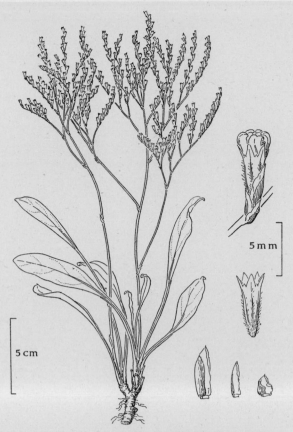

957. *Limonium humile* Mill. Lax-flowered Sea Lavender
Blue-purple

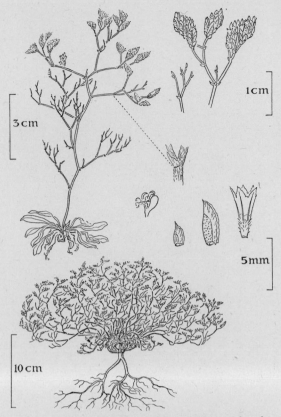

958. *Limonium bellidifolium* (Gouan) Dum. Matted
Sea Lavender Pale lilac

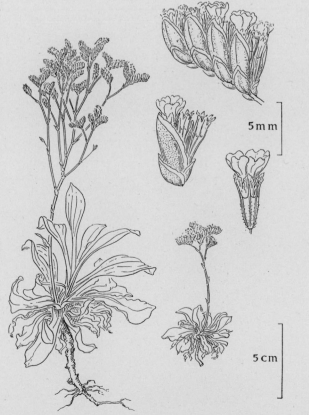

959. *Limonium auriculae-ursifolium* (Pourr.) Druce
Violet-blue

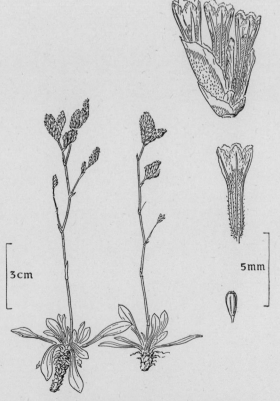

960. *Limonium binervosum* (G. E. Sm.) C. E. Salmon
Rock Sea Lavender Violet-blue

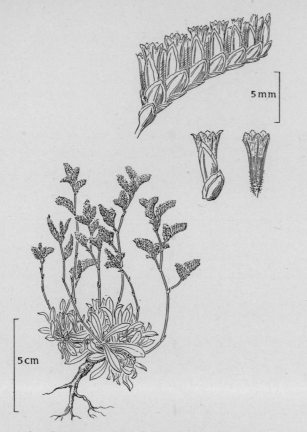

961.   *Limonium recurvum* C. E. Salmon   Violet-blue

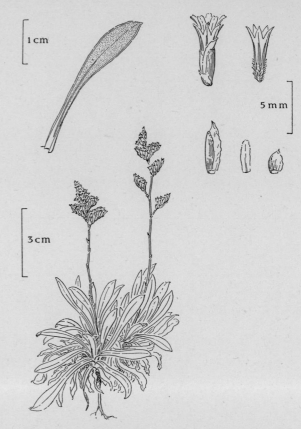

962.   *Limonium transwallianum* (Pugsl.) Pugsl.   Violet-blue

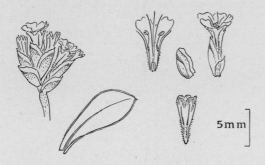

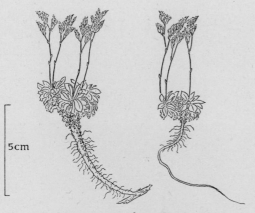

963.   *Limonium paradoxum* Pugsl.   Violet

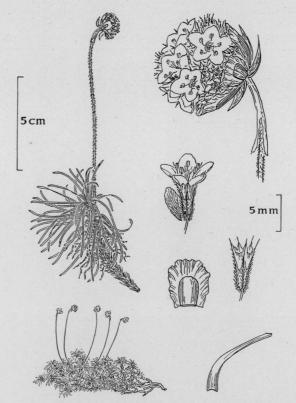

964.   *Armeria maritima* (Mill.) Willd.   Thrift, Sea Pink
Pink or white

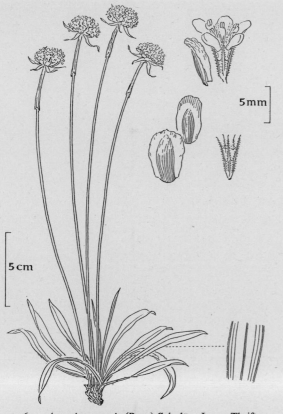

965. *Armeria arenaria* (Pers.) Schult. Jersey Thrift
Deep pink

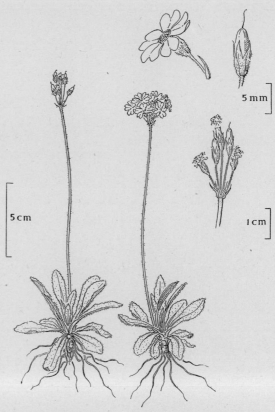

966. *Primula farinosa* L. Bird's-eye Primrose Rosy lilac

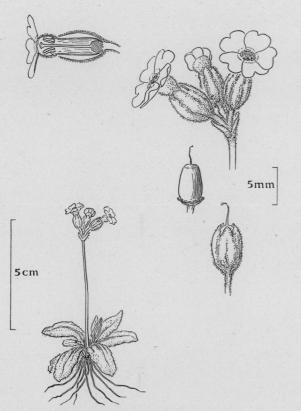

967. *Primula scotica* Hook. Purple

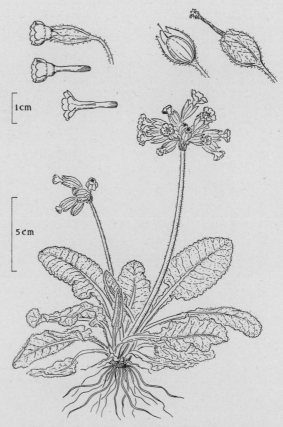

968. *Primula veris* L. Cowslip, Paigle Deep yellow

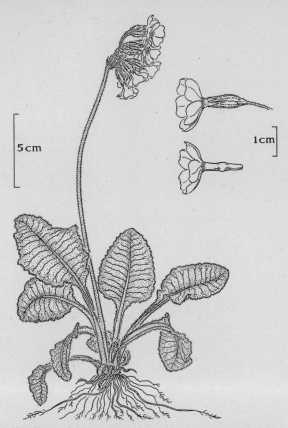

969. *Primula elatior* (L.) Hill   Oxlip, Paigle   Pale yellow

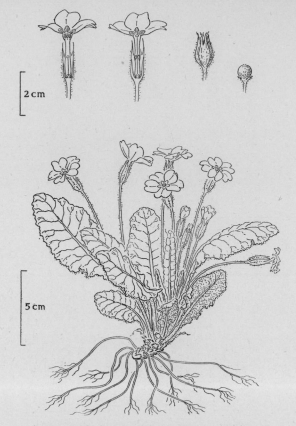

970. *Primula vulgaris* Huds.   Primrose   Pale yellow

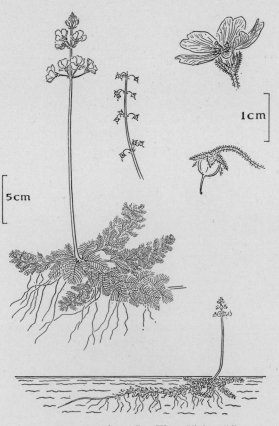

971. *Hottonia palustris* L.   Water Violet   Lilac

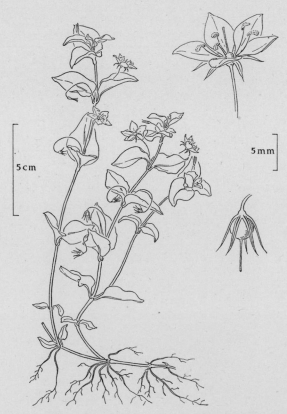

972. *Lysimachia nemorum* L.   Yellow Pimpernel   Yellow

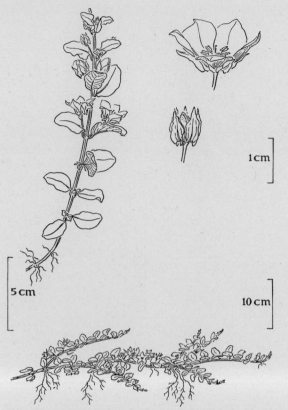

973. *Lysimachia nummularia* L.   Creeping Jenny   Yellow

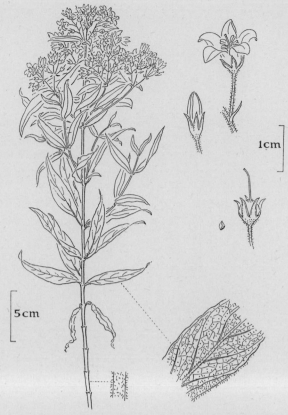

974. *Lysimachia vulgaris* L.   Yellow Loosestrife   Yellow

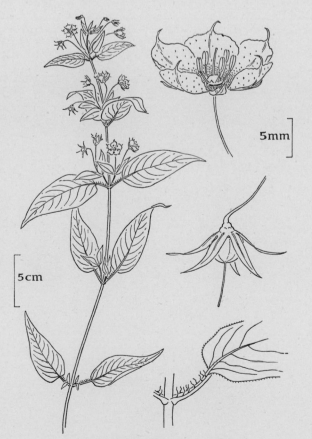

975. *Lysimachia ciliata* L.   Yellow

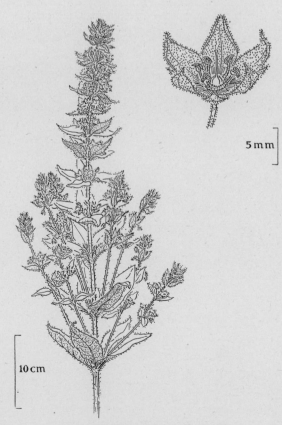

976. *Lysimachia punctata* L.   Yellow

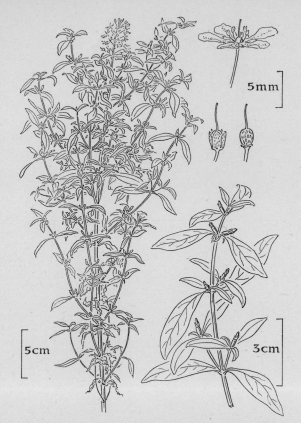

977.   *Lysimachia terrestris* (L.) Britton, Sterns & Poggenb.
Yellow

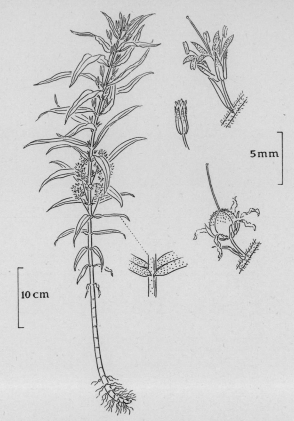

978.   *Naumburgia thyrsiflora* (L.) Rchb.   Tufted Loosestrife
Yellow

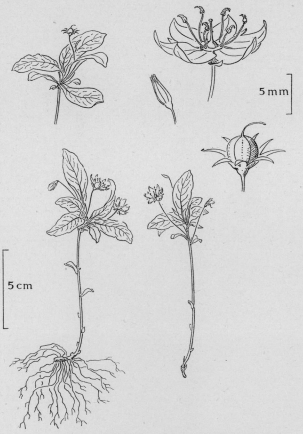

979.   *Trientalis europaea* L.   Chickweed Wintergreen
White

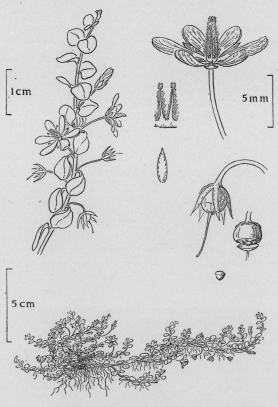

980.   *Anagallis tenella* (L.) L.   Bog Pimpernel   Pink

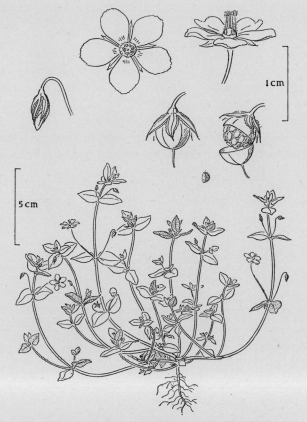

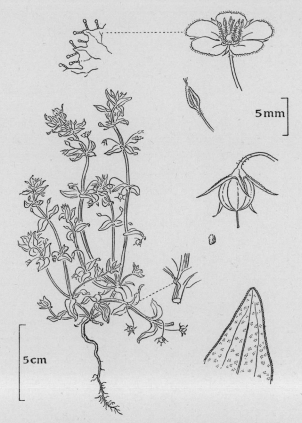

981. *Anagallis arvensis* L. ssp. *arvensis* Scarlet Pimpernel
Red (blue)

982. *Anagallis arvensis* ssp. *foemina* (Mill.) Schinz & Thell.
Blue

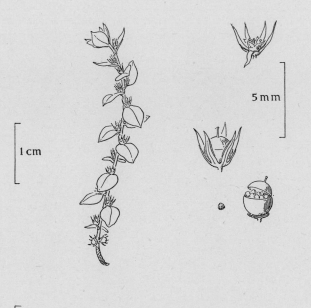

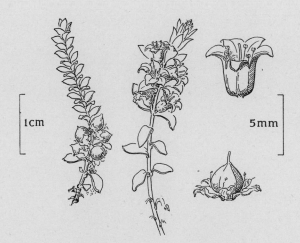

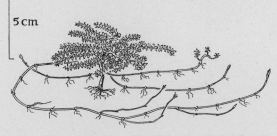

983. *Anagallis minima* (L.) E. H. L. Krause Chaffweed
White or pink

984. *Glaux maritima* L. Sea Milkwort, Black Saltwort
Pink

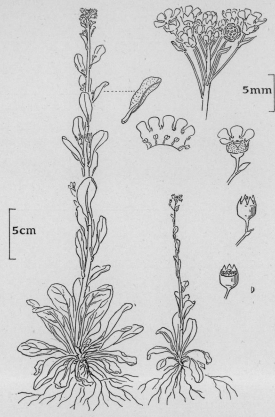

985.   *Samolus valerandi* L.   Brookweed   White

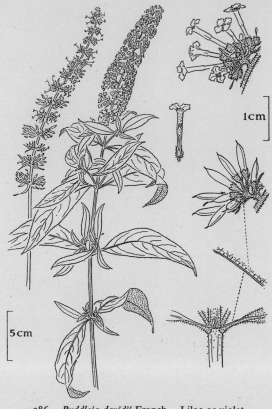

986.   *Buddleja davidii* Franch.   Lilac or violet

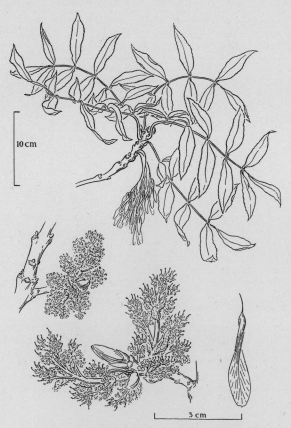

987.   *Fraxinus excelsior* L.   Ash   Purplish

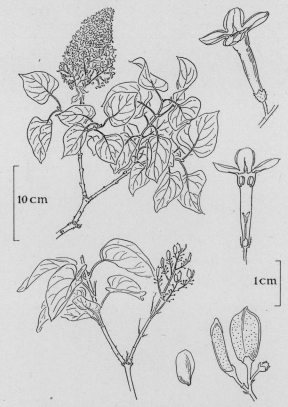

988.   *Syringa vulgaris* L.   Lilac   Lilac or white

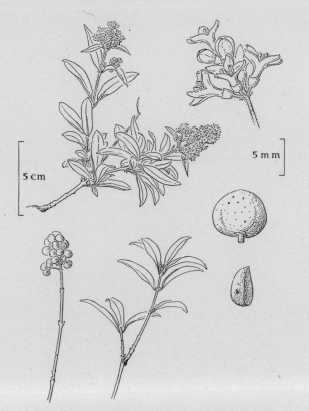

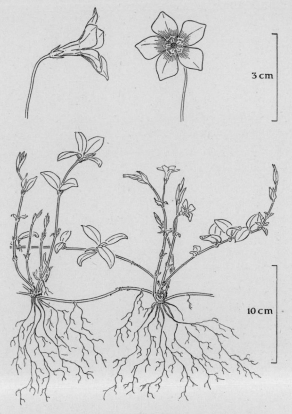

989.   *Ligustrum vulgare* L.   Common Privet   White

990.   *Vinca minor* L.   Lesser Periwinkle   Blue-purple

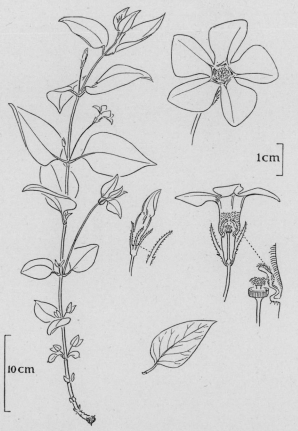

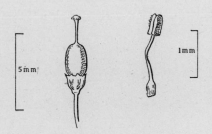

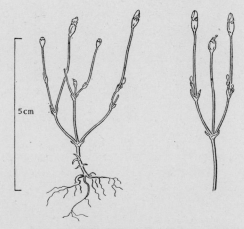

991.   *Vinca major* L.   Greater Periwinkle   Blue-purple

992.   *Cicendia filiformis* (L.) Delarb.   Yellow

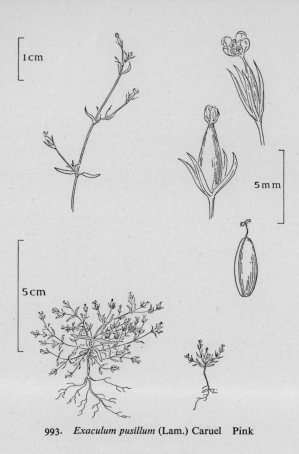

993. *Exaculum pusillum* (Lam.) Caruel   Pink

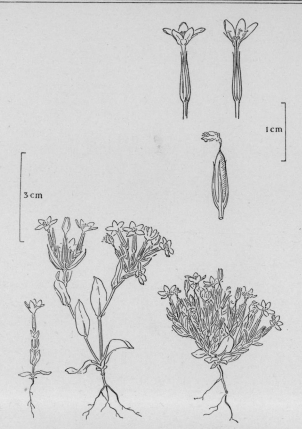

994. *Centaurium pulchellum* (Sw.) Druce   Pink

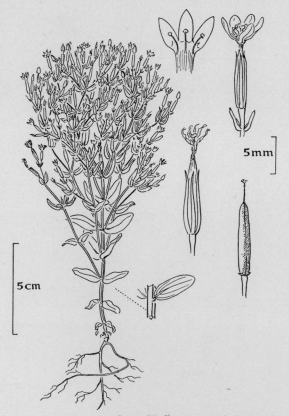

995. *Centaurium tenuiflorum* (Hoffmanns. & Link) Fritsch
Pink

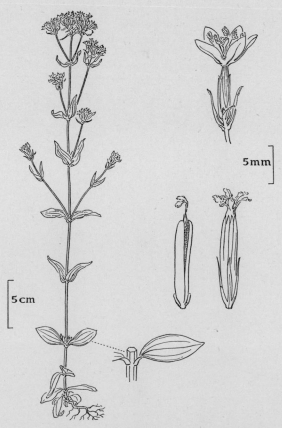

996. *Centaurium erythraea* Rafn   Common Centaury
Pink

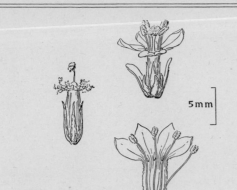

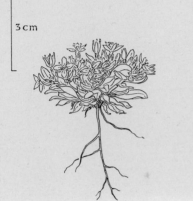

5mm

3cm

997. *Centaurium capitatum* (Willd.) Borbás Pink

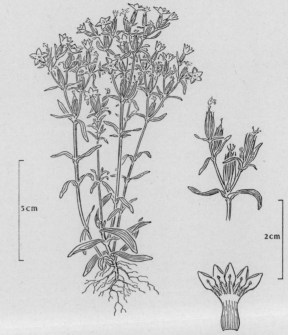

5cm

2cm

998. *Centaurium littorale* (D. Turner) Gilmour Pink

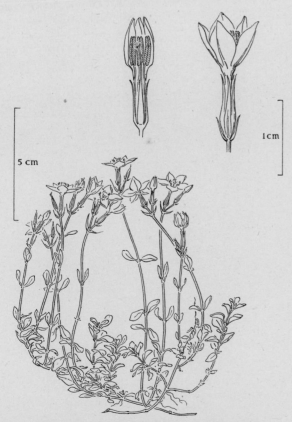

1cm

5 cm

999. *Centaurium portense* (Brot.) Butcher Pink

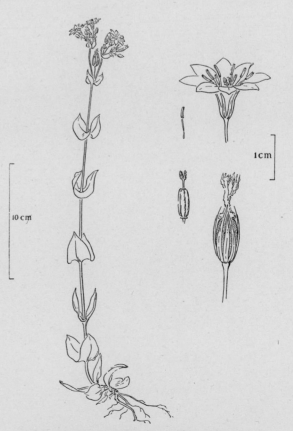

1cm

10 cm

1000. *Blackstonia perfoliata* (L.) Huds. Yellow-wort
Yellow

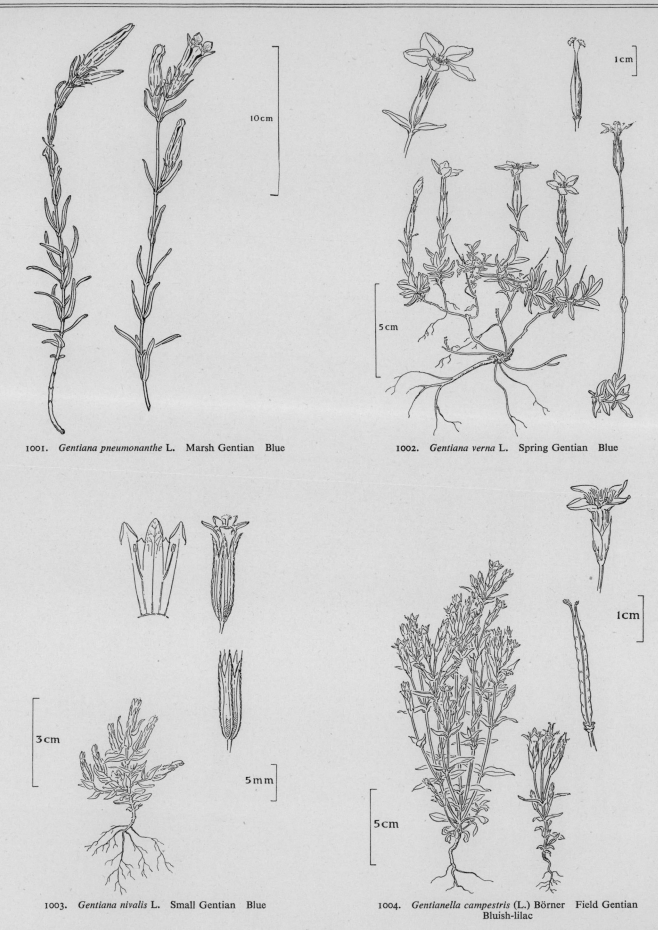

1001. *Gentiana pneumonanthe* L.   Marsh Gentian   Blue

1002. *Gentiana verna* L.   Spring Gentian   Blue

1003. *Gentiana nivalis* L.   Small Gentian   Blue

1004. *Gentianella campestris* (L.) Börner   Field Gentian
Bluish-lilac

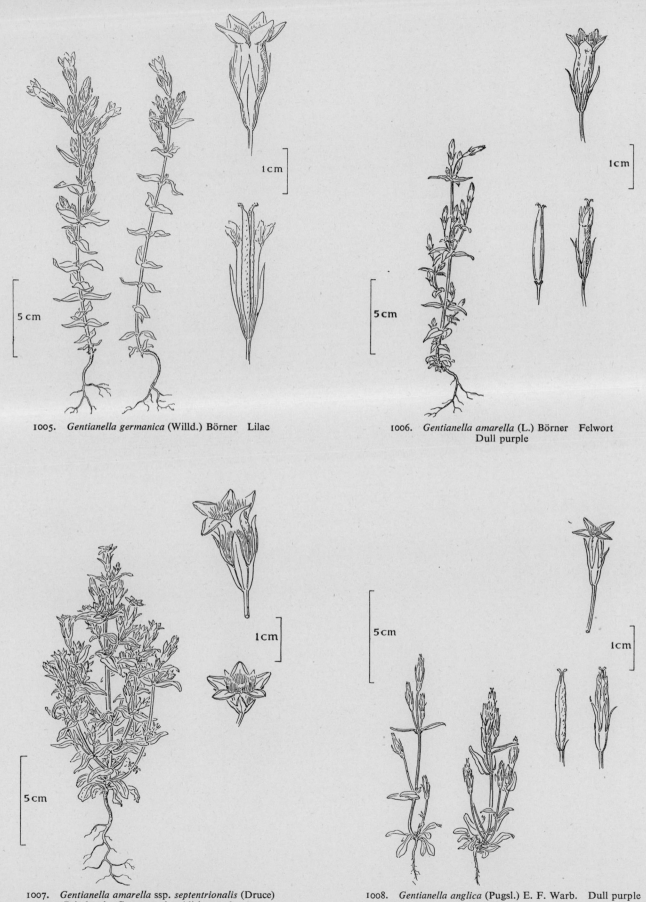

1005.   *Gentianella germanica* (Willd.) Börner   Lilac

1006.   *Gentianella amarella* (L.) Börner   Felwort
Dull purple

1007.   *Gentianella amarella* ssp. *septentrionalis* (Druce)
Pritchard   Cream and reddish-purple

1008.   *Gentianella anglica* (Pugsl.) E. F. Warb.   Dull purple

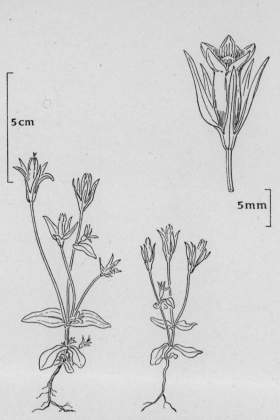

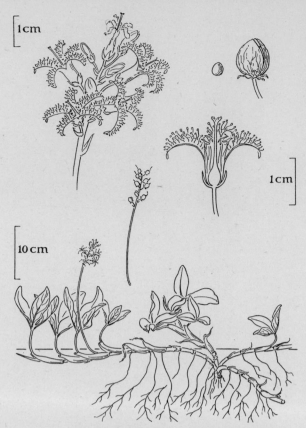

1009. *Gentianella uliginosa* (Willd.) Börner Dull purple

1010. *Menyanthes trifoliata* L. Bogbean, Buckbean
Pinkish-white

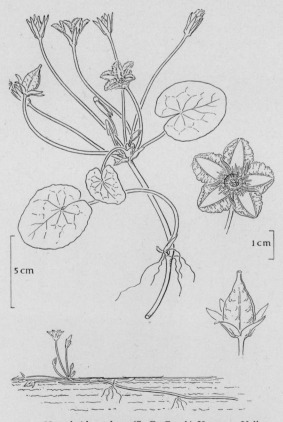

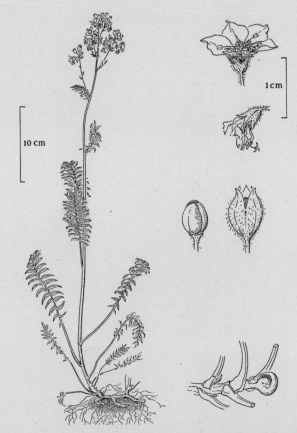

1011. *Nymphoides peltata* (S. G. Gmel.) Kuntze Yellow

1012. *Polemonium caeruleum* L. Jacob's Ladder Blue

INDEX